PLANT FOSSIL ATLAS FROM (PENNSYLVANIAN) CARBONIFEROUS AGE FOUND IN CENTRAL APPALACHIAN COALFIELDS

By

Thomas F. McLoughlin

Geologist, M.S.

© Copyright 2016 Thomas McLoughlin.

All rights reserved. No part of this publication may be reproduced, stored in a retrieval system, or transmitted, in any form or by any means, electronic, mechanical, photocopying, recording, or otherwise, without the written prior permission of the author.

Created in the United States of America.

Because of the dynamic nature of the Internet, any web addresses or links contained in this book may have changed since publication and may no longer be valid. The views expressed in this work are solely those of the author and do not necessarily reflect the views of the publisher, and the publisher hereby disclaims any responsibility for them.

Any people depicted in stock imagery provided by Thinkstock are models, and such images are being used for illustrative purposes only. Certain stock imagery © Thinkstock.

Green Ivy Publishing
1 Lincoln Centre
18W140 Butterfield Road
Suite 1500
Oakbrook Terrace IL 60181-4843
www.greenivybooks.com

ISBN: 978-1-945650-58-1

Pennsylvanian coal swamp vegetation reconstruction, a composite of many plant types growing in and around the swamp (Kukuk, 1938).

ACKNOWLEDGEMENT

This book could not have been completed without the dedicated help of Cortland F. Eble, Ph.D., and Alton Dooley, who are paleontologists with the Kentucky Geological Survey in Lexington, Kentucky, and the Museum of Natural History in Martinsville, Virginia, respectively. They helped edit the manuscript. Assistance in the classification of many of the fern fossils was given by Dr. Shusheng Hu, who is a paleobotanist and Collections Manager, Division of Paleobotany at the Yale Peabody Museum of Natural History in New Haven, Connecticut.

I also want to thank my wife, Beth, for her patience and tolerance for the numerous boxes of fossil specimens in our home. She was very relieved when I donated the collection to the Museum of Natural History.

All of the fossils listed in the plates were collected by and photographed by the author except as noted.

FOREWORD

I have spent the last twenty-seven-plus years in and around the bituminous coal mines of southwestern Virginia. When coal miners learn I am a geologist, the most popular question has been "what are the kinds of fossils we see in a mine roof?" I give my best reply, but it is difficult to relate to them that the plant impressions represent vegetation that grew in peat-forming swamps millions of years ago. Most people recognize the fern-like fossils, but have been confused about the identity of a portion of tree root versus the tree itself. Many believe that the fossils are not those of ancient vegetation, but instead are the preserved remains of fish or reptiles.

I became interested in geology because of these fossils. It is the goal of this publication to share my accumulated experience in the area of basic paleobotany and furnish a pictorial guide to the identification of the more common Carboniferous-age plant fossils from the coal fields of Virginia. Those especially targeted are the rock hounds and aspiring geologists of all ages.

In 1977, I received my Bachelor of science degree from Morehead State University (MSU) in Morehead, Kentucky. In the spring of 1980, I graduated from Eastern Kentucky University (EKU) in Richmond, Kentucky, with a Master of science degree in geology.

During those years, the majority of my geologic experiences centered on the geologic aspects of underground coal mine roof stability by benefit of U.S. Bureau of Mines contracts awarded to a professor at MSU, Dr. David K. Hylbert. I owe a large part of my success as a geologist to Dr. Hylbert; Dr. Harry Hoge, my thesis adviser at EKU; and Dr. Jules DuBar, my paleontology professor while I was at MSU. Therefore, I wish to dedicate this publication to them as thanks for their guidance and inspiration.

INTRODUCTION

Fossils have excited people for a long time, but for about 400 years, the term was used to describe almost anything that looked like it had organic origins and was dug up from the earth. "Fossil" is defined by paleontologists as any object that represents the presence of a former life, as the term also applies to the preservation of various trace fossils such as animal trackways and coprolites (fecal pellets). By convention, use of the term is generally restricted to remains that are older than 10,000 years.

The study of fossilized plant remains is called paleobotany. Understanding how plants inhabited the earth throughout geologic time allows the paleobotanist to begin to piece together the history of the plant kingdom. Fossil plants come in a variety of shapes and sizes that vary throughout geologic time. Examining and Identifying species that lived millions of years ago allows us to glimpse into ecological, and therefore, evolutionary occurrences. Generally, the preservation of an organism requires a rapid burial in sediment, usually clay (mud), silt, or fine grained sand, before the soft body portions completely decay or are fragmented to such an extent that it cannot be identified as a specific type of organism. Even after preservation, few fossils are discovered and collected before weathering and erosion destroy the rocks that carry them.

Fossil plants can be preserved in a variety of ways. Most commonly, the shape of the plant is impressed into the sediment. During this process, plant material falls into the water, becomes water-logged, sinks to the bottom of the body of water, and becomes surrounded and covered by sediments. Slowly, under the increasing weight of the additional sediments, water and air are pressed out until only plant material remains. The flattened plant part appears as a fossil compression on one layer of the strata, while the other side contains the impressed counterpart or "impression."

Frequently root systems, trunks, and limbs in the proper growing position of plants become engulfed by sediments during floods when streams and rivers overflow their banks or shift their courses. Sediment partially or completely replaces decaying plant (organic) material so that the walls of the resulting cavity (or mold) form with exact details. Standing tree or trunk casts that were buried in this fashion are called "kettle bottoms" or "stove pipes" by the mining industry because of the flared or bowl shape at the base and upward taper.

Often as the depth of burial increases, heat and pressure builds causing the gradual loss of original organic tissue to the extent that only a layer of carbonaceous material (coal) remains, a process known as carbonization. Often these fossils are the most spectacular and "pretty," since even the most delicate details of leaves, barks of trees, and branches are preserved in an almost life-like fashion. Veins and filaments stand out in stark relief in these fossil examples. This is the typical type of preservation found in coal seams.

If the tree became buried in sediment and water percolated through the ground, then each individual cell of the organism might be replaced by dissolved minerals including silica (quartz or jasper), calcium carbonate (calcite), or iron magnesium carbonate (ironstone). This would result in petrifaction of the organism.

The Appalachian region of the United States is full of plant fossils that represent a thriving ecosystem during the Carboniferous period (360 to 286 million years ago); this guide focuses on the most wide-spread and commonly found flora of that period. Unique water chemistry and a tropical climate created extensive coal swamps in the Appalachian Mountains; under these conditions it is very rare to find the hard parts of animals because they were not preserved well. Some shell fossils of brackish water to shallow marine brachiopods are sometimes locally abundant in certain rock units (e.g., the Magoffin Beds in the Wise Formation) in the tristate region (Virginia, West Virginia, and Kentucky). A few brachiopods, pelecyepods, and nautiloids associated with the plant fossils were also collected, but these are very small and easily overlooked by the untrained eye. Even though there are numerous plant fossils to be found, these most likely represent only a small fraction of the abundant flora that existed because plants are so susceptible to decay.

The majority of the specimens pictured in this publication were collected from coal mines in southwestern Virginia. Of the numerous coal seams that are mined in Virginia, there are a few that had special conditions conducive to optimum preservation. These include the Jawbone, Lower Banner, Upper Banner, Splashdam, Kennedy, Hagy, and Taggart seams; all of these seams are Pennsylvanian in age (a standard geologic time scale is shown in figure 1). The locations of the collection sites are listed in Appendix A.

Throughout this reference, there are comparisons made between fossil and present-day plants to aid in the interpretation of structures observed in the extinct plants. Several plants found living today are also preserved as fossils and show few little changes in morphology (appearance). Most notable is the group of extinct plants called Calamites; the modern horsetail is closely related to Calamites, but horsetails are much smaller in size. Because of this, the horsetail is often referred to as a "living fossil."

A most notable collecting site is one located just west of Coeburn, Virginia, along the westbound stretch of U.S. Route 58 Alternate. The section represents an ancient volcanic ash bed as evidenced by an unusually high amount of the clay mineral montmorillonite, which is known to be a weathering product of volcanic ash. The clay is also mixed in with the sandstone. Together with the sandstone and clay, there are layers of rock at least 50 feet in thickness. Specimens of *Lepidodendron*, *Sigillaria*, *Calamites*, seed ferns *Neuropteris* and *Alethopteris*, pollen organs *Whittleseya*, and a seed pod *Holcospermum* were found at this location near Coeburn in Wise County, Virginia, immediately above the Aily coal seam.

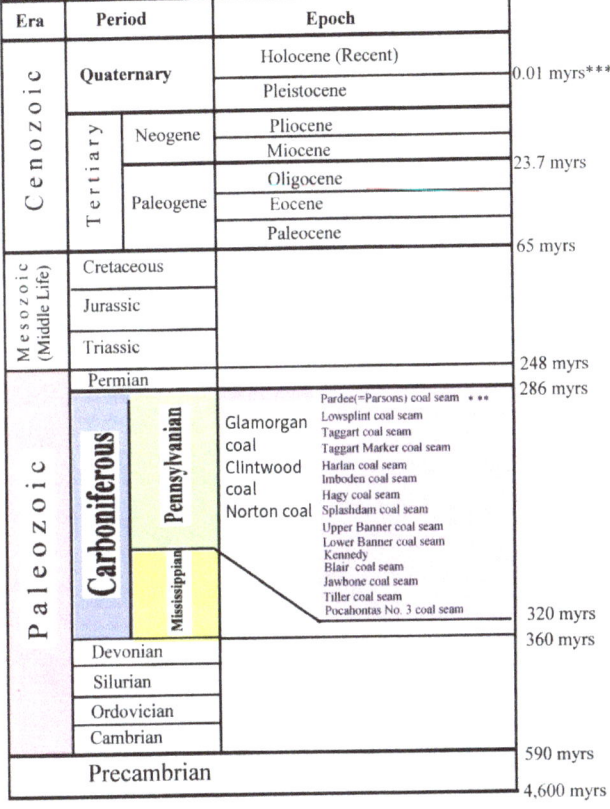

* Modified from Geologic Time Scale posted on the United States Geological Survey Web site.

** Millions of Years Before Present

*** Relative stratigraphic positions of coal horizons in Southwestern Virginia from which plant fossils have been collected.

Figure 1
General geologic time scale showing the various coal seams from which fossils were collected. Also see Appendix A for names of the formations in which the fossils were found.

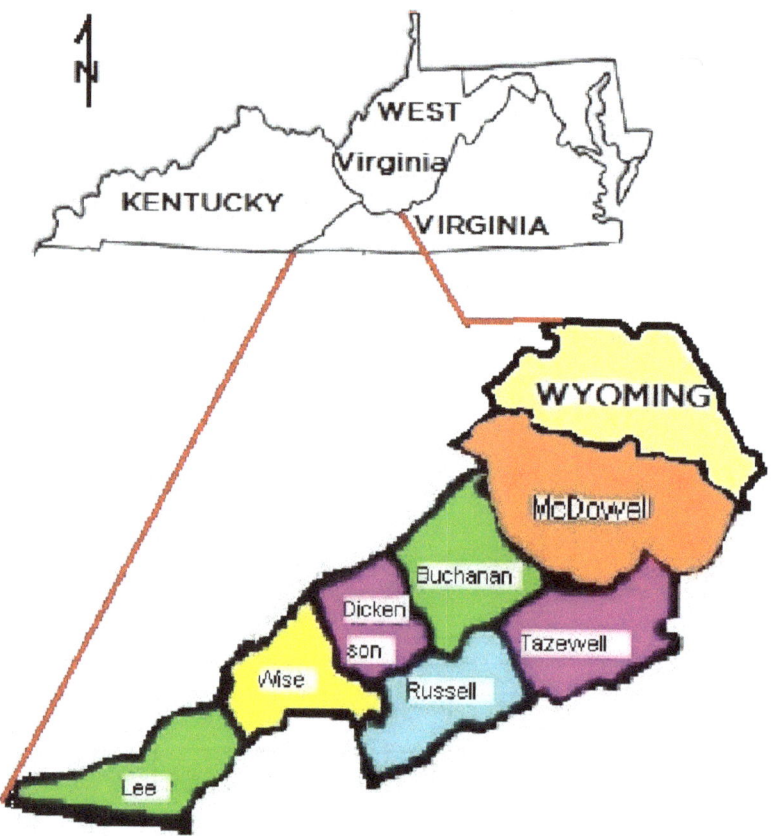

Figure 2. Index map of the study area where fossil flora have been collected from southwestern Virginia and southwestern, West Virginia. Appendix A list the more detailed description of the precise location of the fossil collection sites.

COLLECTING PLANT FOSSILS

To the beginner, finding plant fossils or fossils in general can be frustrating. Often the first time out or even the second may not be productive. However, I have learned that perseverance and patience will win out. There is also a certain degree of luck involved.

You need to get permission from the landowner before you enter and collect. Do not cross locked gates or posted "No Trespassing" signs. Although it may be tempting, abandoned coal mines are always prohibited from entry.

The basic tools required for successful collecting are as follows: (1) A masons' hammer—one with a chisel (wide blade) end, (2) a handheld rock splitter for larger rocks, (3) gloves, (4) safety goggles or other eye protection, (5) safety shoes or hard-toe boots, and (6) a hard hat (see figure 3). Always wear a hard hat and proper footwear to protect you from falling debris. Remember safety first. Lists of fossil-collecting localities are generally not published. In planning a field trip, you can find your own sites by learning to use geologic and topographic quadrangle maps for your area. They contain an abundance of information useful to the collector.

Plant fossils are usually found at or immediately above a coal seam. Once you find one, first conduct a quick survey of the talus (i.e., weathered material—fragments of rock that accumulate at the base of an outcrop or road cuts).

Once a locality proves to have the potential for bearing fossils, it is time to start breaking rocks. Remember to don your safety equipment before commencing the hunt. This is where patience is required because it will require the splitting and resplitting of many rocks before the fruits of your labor are realized. In fact, many times the strata will be so deeply weathered that only fragments of what would otherwise have been a whole specimen come out as small pieces or crumble in your hands. Thus some digging is required to get to the firmer rock. Often it is necessary to migrate laterally and vertically in the outcrop to find a horizon that is fossil bearing. Based on experience, the best advice is to follow the fossils. Just like in the movie *The Wizard of Oz*, once on the "yellow brick road," one continues upward from the talus in search of the source of the specimen. Also, be alert to the encounter of marine-type fossils such as snails and clams, as described earlier in the introduction. You may discover a fossil which may have eluded the trained eye of the professional geologist.

In transporting the specimens home, be sure to take precautions against the potentially damaging effects of excessive vibration in a vehicle. Be sure to use plenty of padding regardless of the size and firmness of the rock. After arriving home, the next step is to clean your specimens. Use a very fine and soft hairbrush (most recommended is one made of camels' hair) to clear away loose debris. Ask your dentist for instruments that are to be discarded; inform him that they make excellent tools for breaking away thin surface layers of rock to more fully expose plant fossils.

Next, photograph your prizes to share with others (some even make great wallpaper for a computer screen). Should the contrast be very small between the color of the fossil and its matrix, a very thin coating of polyurethane may be required to bring out the full beauty of the specimen. This process will also retard the rate of decay of the rock and protect against any major damage to the fossil. Most are preserved in a thin film of carbon, which may flake and peel over time, especially when handled frequently.

Using either acrylic paint or correction fluid, make a large enough area to document the location from which the sample was collected. Someone else may want to visit the site in the hopes of finding their own. An identification card with additional information is often included in the same container or place in which you store or display the fossil.

Identification of the fossils with specific names or classifications takes a lot of practice. The best way to start is to go to a library and find reference books. Most have photographs and drawings of the most common plant fossils that you can compare with your own. To help verify the name of the fossil, read the descriptions offered by the books. Conferring with a geologist (and better, a paleontologist) is usually needed for identifying the genus and species. Browsing the Internet (as I did) can also offer very valuable information.

Figure 3. Safety equipment and tools for collecting fossils. The camel-hair brush and old dentist's tools are used to clean and dress the fossils in preparation for a protective sealant and photographing.

Table of Contents

ACKNOWLEDGEMENT IV

FOREWORD V

INTRODUCTION VI

COLLECTING PLANT FOSSILS XI

CHAPTER
    ARBORESCENT LYCOPODS (CLUB MOSSES, SCALE TREES)
    Lepidodendron 1

CHAPTER 2
    ARBORESCENT LYCOPODS (CLUB MOSSES)
    Sigillaria 19

CHAPTER 3
    CORDAITES: EARLY GYMNOSPERMS
    Ancient Mangrove-Like Plant 28

CHAPTER 4
KETTLEBOTTOMS
    Ancient Tree Trunks...34

CHAPTER 5

STIGMARIA

    Ancient Root Systems 39

CHAPTER 6

    CALAMITES

    Ancient Relative of the "Horsetail" 46

CHAPTER 7

    SPHENOPHYLLUM 75

CHAPTER 8

    FERNS 82

CHAPTER 9

    SEEDS 130

CHAPTER 10

    MARINE FOSSIL FAUNA FOUND WITH PLANT FOSSIL FLORA 137

REFERENCES: 144

APPENDIX A 146

CHAPTER 1

ARBORESCENT LYCOPODS (CLUB MOSSES, SCALE TREES)

Lepidodendron

Lepidodendron. This plant is sometimes referred to as a "scale tree" because of the distinctive teardrop or diamond-shaped pattern of the bark of this lycopod. It is often mistaken for the scales of a reptile or a snake's skin. Each scale-like feature is accented by a small depression that looks like an eye. The modern relatives of this plant are the ground pine, running cedar, and club moss, or *Lycopodium* (see figure 4a). As a result, *Lepidodendron* has been described as a giant club moss. Note that neither *Lepidodendron* nor *Lycopodium* is a pine, a cedar, or a moss. The scar morphology of the branches appears as a miniature version of scars found on the main stem. They appear as oval-shaped depressions from one to two inches in size. Others appear as nearly circular, dome-shaped features recessed into the bark. These are referred to as *Ulodendron* (see plate I, *Ulodendron)*. It is also thought by some scientists that these shallow teardrop-shaped features may represent the points of attachment of reproductive cones or pods. These plants generally stood as high as 98 feet and were common during the Carboniferous period.

Ground pine and club moss are the common names for a small terrestrial evergreen that looks like a miniature pine tree with small scaly leaves that grow in patches (see figure 4). The genus *Lycopodium* and other members of the *Lycophyta* (club mosses) have their origins in the Carboniferous as giant trees (e.g., *Lepidodendron*) (see figure 4).

Figures 4a and 4b are examples of ground pines (club mosses) with reproductive cones (arrows). Lycopodium sp. (left) and Lycops (right) are living descendants of Lepidodendron. Figure 4c is a reconstruction of Lepidodendron, with reproductive organs shown in orange. Modified after Gillespie et al., 1978.

Before the anatomy of *Lepidodendron* (i.e., the microscopic examination of its tissue cells) was understood, generic names for the plant stems were based on the various appearances of the same plant form resulting from preservation at different stages of decortication, or states of decay. The generic terms, which include *Knorria, Bergeria,* and *Aspidiaria*, have been retained for descriptive purposes only (see figure 5). These forms of stem casts are more fully discussed by Seward (1898). Two of these are illustrated in plate I (*Lepidodendron* numbers 2, 3, and 4).

Species of *Lepidodendron* are based on the pattern and morphology of the leaf cushions on the surface (bark). *Sigillaria* and *Lepidodendron* are differentiated by the arrangement of the leaf cushions and scars. *Sigillaria* leaves were formed in vertical columns in contrast to *Lepidodendron*, which were spirally disbursed along the stems. The distinguishing feature of a well-preserved leaf-cushion of the genus *Lepidodendron* is a rhomboidal or fusiform cushion that is elongated longitudinally, somewhat reminiscent of scales found on fish and reptiles. Thus, the term "scale tree" has been associated with *Lepidodendron*. The leaf-scar, or place of attachment on the base of the leaf, is in turn a clearly defined smooth area located in the middle portion of the leaf-cushion. The anatomical characteristics for naming the *Lepidodendron* species are presented in plate II (*Lepidodendron*), numbers 1, 6, and 6a, and plate III (*Lepidodendron*), numbers 2a and 3a.

Not all of the species of *Lepidodendron* have been found as of the date of this publication. However, all coal seam horizons have not been visited. Future studies may unearth additional forms of *Lepidodendron*, as well as other Paleozoic flora.

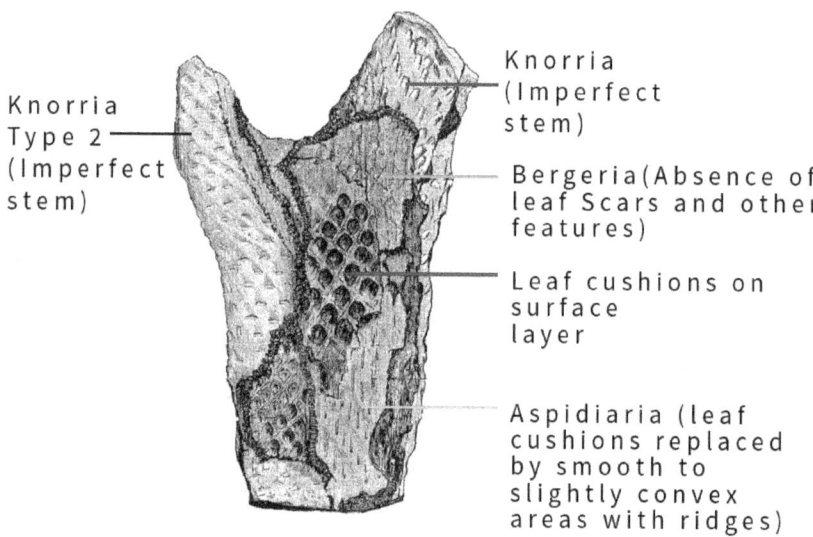

Figure 5. Illustration of the different stages of decortication, or states of decay, using a specimen of *Lepidodendron*. Modified after Seward, 1889, vol. II, fig. 156, p. 125.

Plate I. 1. *Lepidodendron aculeatum* preserved in sandstone. Collected from the strata directly overlying the Kennedy coal seam along U.S. Route 58 Alternate North, approximately 1.5 miles east of Coeburn in Wise County, Virginia.

2. *Lepidodendron veltheimianum* in the Aspidiaria stage of decortication preserved in shale.

3. *Lepidodendron veltheimianum* Aspidiaria stage 4. Collected from the roof strata of a mine in the Parsons coal seam along Mud Lick Creek, north of Roda in Wise County, Virginia.

4. *Lepidodendron veltheimianum* in the Knorria stage of decortication preserved in shale. Collected from the roof rock of the Parsons coal seam near Roda in Wise County, Virginia, on Mud Lick Creek. Collected from the roof strata of a mine in the Upper Banner coal seam near Bucu in Dickenson County, Virginia.

5. *Lepidodendron veltheimianum* preserved in shale. Collected from the roof strata of a mine where the Lower Banner and Splashdam coal seams merged into one seam, northwest of Coeburn in Wise County, Virginia.

6, 6a. *Lepidodendron obovatum* preserved in shale. Collected from the roof strata of a mine in the Taggart Marker coal seam in Stonega, Wise County, Virginia.

7. *Lepidodendron halonia.* 7a Reverse side. Collected from the roof strata of a mine in the Parsons coal seam along Mud Lick Creek, north of Roda in Wise County, Virginia.

Plate II. 1. *Bothrodendron* cf. *B. punctatum* inside yellow box. 1a. Enlarged view of the morphology below the outer layer of the area enclosed by the red box. Collected from the roof strata in the Parsons coal seam along Mud Lick Creek, north of Roda in Wise County, Virginia.

2. *Lepidodendron obovatum.* Enlarged view of surface morphology.

Collected from the roof strata in a mine in the Splashdam coal seam along Abners Fork (State Route 670) near Hurley in Buchanan County, Virginia.

3. *Lepidodendron obovatum*. 3a. Enlarged view of surface morphology. Collected from the roof strata in a mine in the Tiller coal seam west of U.S. Route 460 North near Shortt Gap, Buchanan County, Virginia.

4. *Lepidodendron obovatum*. 4a. Enlarged view of surface morphology. Collected from the roof strata in a mine in the Pocahontas No. 3 coal seam at the head of Cucumber Creek, 7.7 miles northeast of Squire in McDowell County, West Virginia.

Plate III. 1, 1a, *Lepidodendron obovatum* in a gray clay shale. Collected from an outcrop of the Upper Banner coal seam in a road cut located along I-80 North near Haysi in Dickenson County, Virginia.

2. *Lepidodendron aculeatum* in a medium-grained sandstone. Enlarged view of surface morphology. Collected from above the Aily coal seam approximately 50 feet right off the westbound lane of U.S. Route 58 Alternate, 1.5 miles from Coeburn in Wise County, Virginia.

3. *Lepidodendron aculeatum* in a very fine-grained carbonaceous sandstone. Collected from above the Imboden coal seam in Appalachia, Wise County, Virginia. The specimen was given to me by the collector, Mark Hughes of Pennington Gap, Virginia. He gave me permission to include it in this book.

4, 4a. *Lepidodendron aculeatum* in a grayish-black silty shale. Collected from above the Tiller coal seam in Whitewood/Jewell Ridge, Buchanan County, Virginia.

Plate I Lepidodendron

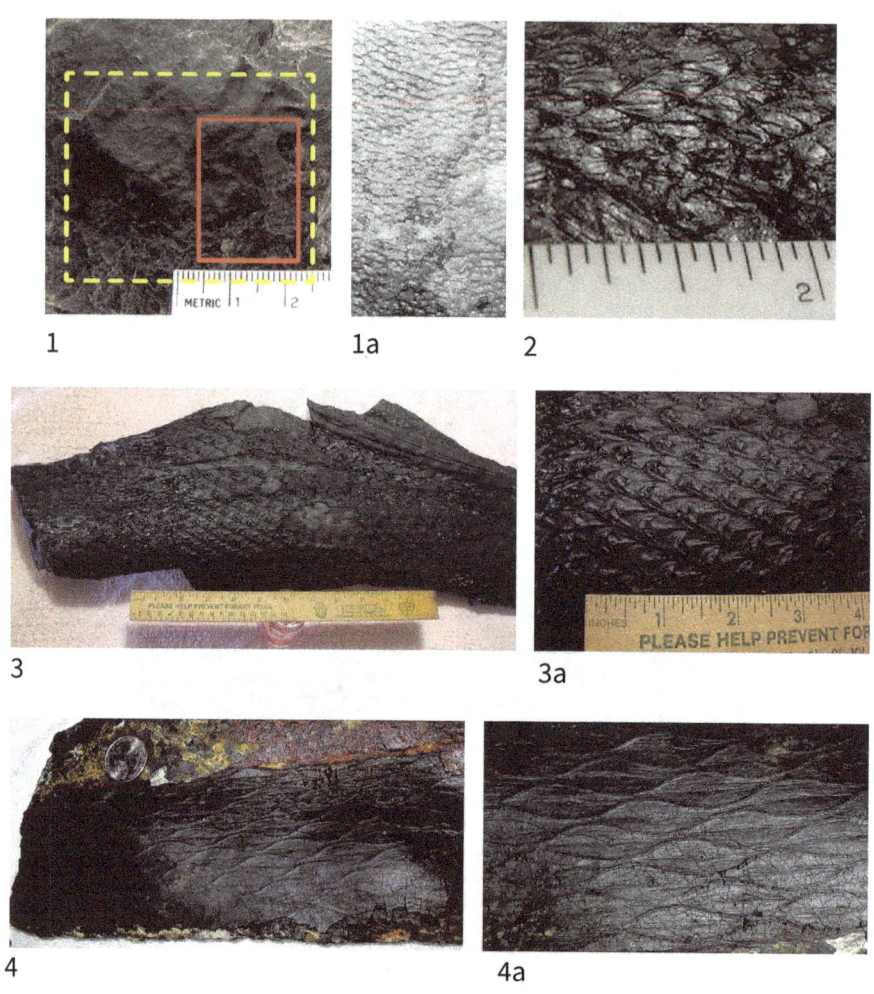

Plate II Lepidodendron

Plate III Lepidodendron

Plate I. 1. *Ulodendron** preserved in shale displaying two well-defined branch scars. 1a. Enlarged portion of branch scar displaying the attachment point of the structure, which is believed to have delivered nutrients to the branch. 1b. A portion of the outer layer greatly magnified to illustrate the characteristic honeycomb-like or scale-like pattern resembling the skin of a fish or reptile. Collected from the roof rocks in a mine developed in the Taggart coal seam in Appalachia, Wise County, Virginia. 2. The shale mold of an *Ulodendron* branch scar.

3. *Ulodendron majus* preserved in a gray shale with several well-defined branch scars. 3a. Enlarged portion of a branch scar and the outer layer to show the surface texture. Collected from the Pocahontas No. 3 coal seam near Vansant in Buchanan County, Virginia, by a coal miner, Zachariah Edwards of St. Paul, Virginia. I was given permission to take photographs and include them in this publication.

* *Ulodendron* is the genus name for a species of fossil tree (lycopod) stem that was once thought to be a part of *Lepidodendron* (Thomas, 1968).

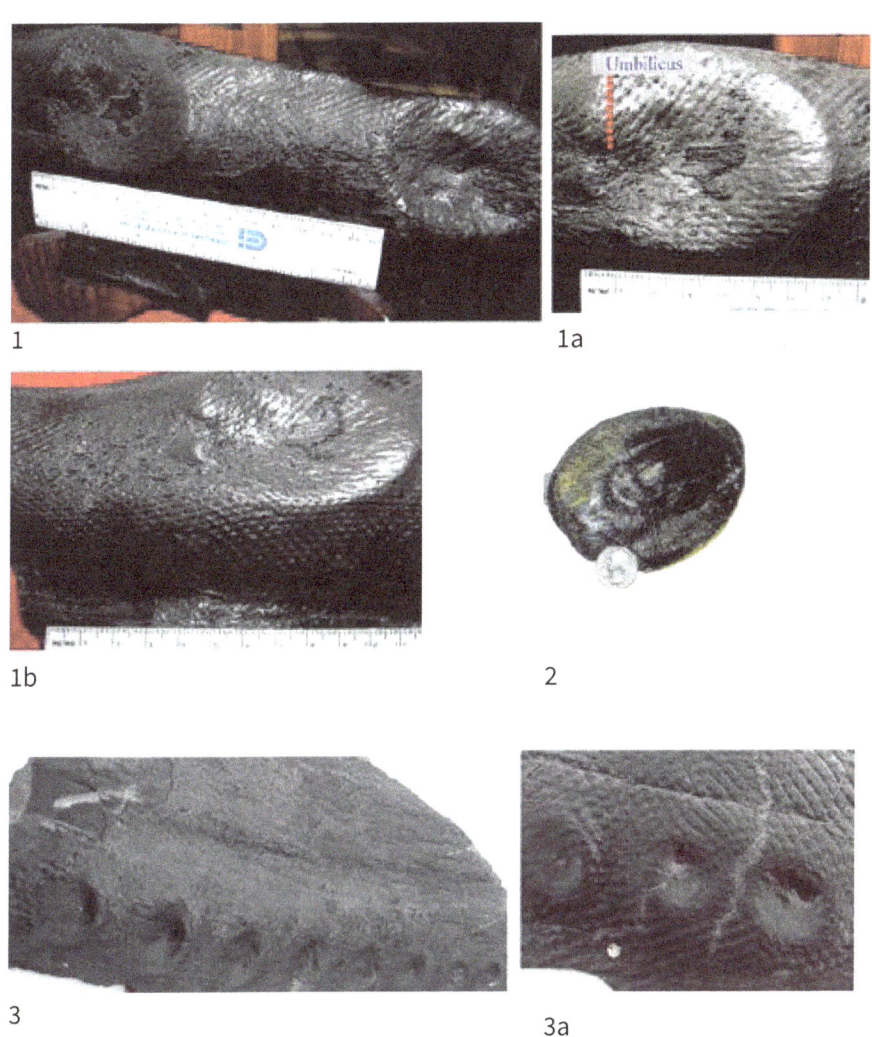

Plate I Ulodendron

Plate I. *Lepidodendron* and *Sigillaria* reproductive organs. 1. *Lepidophlois* sp. in shale. Collected from immediately above the Blair coal seam in the Wise Formation along U.S. Route 58 Alternate, approximately 0.5 miles from Appalachia High School in Appalachia, Wise County, Virginia.

2. *Lepidophyllum* sp. in shale. Collected from the roof strata of the Parsons coal seam along Mud Lick Creek, 2.4 miles northeast of Roda in Wise County, Virginia.

3. *Lepidostrobus* sp. in shale. Collected from the roof strata of a mine in the Lowsplint coal seam, 2.4 miles north of Stonega on Stonega Road, State Route 78, in Wise County, Virginia.

4. A pair of *Lepidostrobus* sp. in shale. Collected from the roof strata of the Parsons coal seam along Mud Lick Creek, 2.4 miles northeast of Roda in Wise County, Virginia.

5. *Lepidostrobus* sp. cast. 6. *Lepidostrobus* sp. mold. Both preserved in shale. Collected from the roof strata of a mine in the Splashdam coal seam located along Smith Branch (State Route 701), 0.7 miles north off Slate Creek (State Route 83), 8 miles east of Grundy in Buchanan County, Virginia.

7. *Sigillariastrobus* Schimper Feistmante in shale. Reproductive cone of *Sigillaria*. Unlike those of *Lepidodendron*, *Sigllariastrobus* grew in clusters and were attached farther back on the branches and not the very tips. Collected from the roof strata of a mine in the Lowsplint coal seam, 2.4 miles north of Stonega on Stonega Road (State Route 78) in Wise County, Virginia.

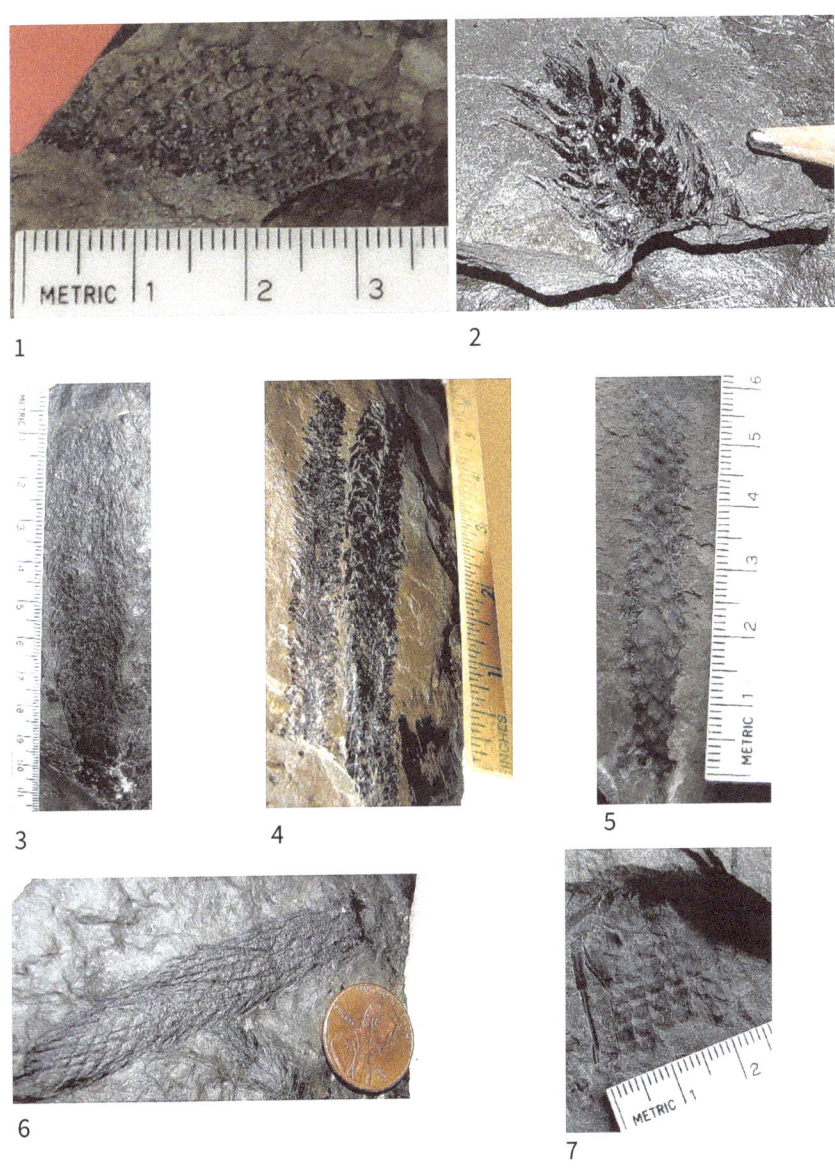

Plate I Lepidodendron & Sigillaria Reproductive organs

Plate I. *Lepidodendron* branches. 1. *Dicranophyllum domini* preserved in shale. 2, 2a. *Dicranophyllum* sp. preserved in shale. The grass-like foliage is still attached to the tree branch. Collected from the roof strata of a mine in the Splashdam coal seam, located along Abners Fork (State Route 670) southeast off State Route 645 near Hurley in Buchanan County, Virginia.

3, 3b. *Lepidodendron sternbergii* . The arrows point to the leaf scars and the linear leaves. 3a. Drawing of leave cushions with linear leaves attached after Seward, 1898, vol. II, fig. 141, p. 97. Collected from the roof strata of a mine in the Lowsplint coal seam, 2.4 miles north of Stonega on Stonega Road, State Route 78, in Wise County, Virginia.

4, 5. *Lepidophylloides* preserved in carbonaceous shale. These specimens illustrate the basis for referring to *Lepidodendron* as the "scale" tree. Collected from the roof strata of a mine in the Parsons coal seam, located along Mud Lick Creek, 2.4 miles northeast of Roda in Wise County, Virginia.

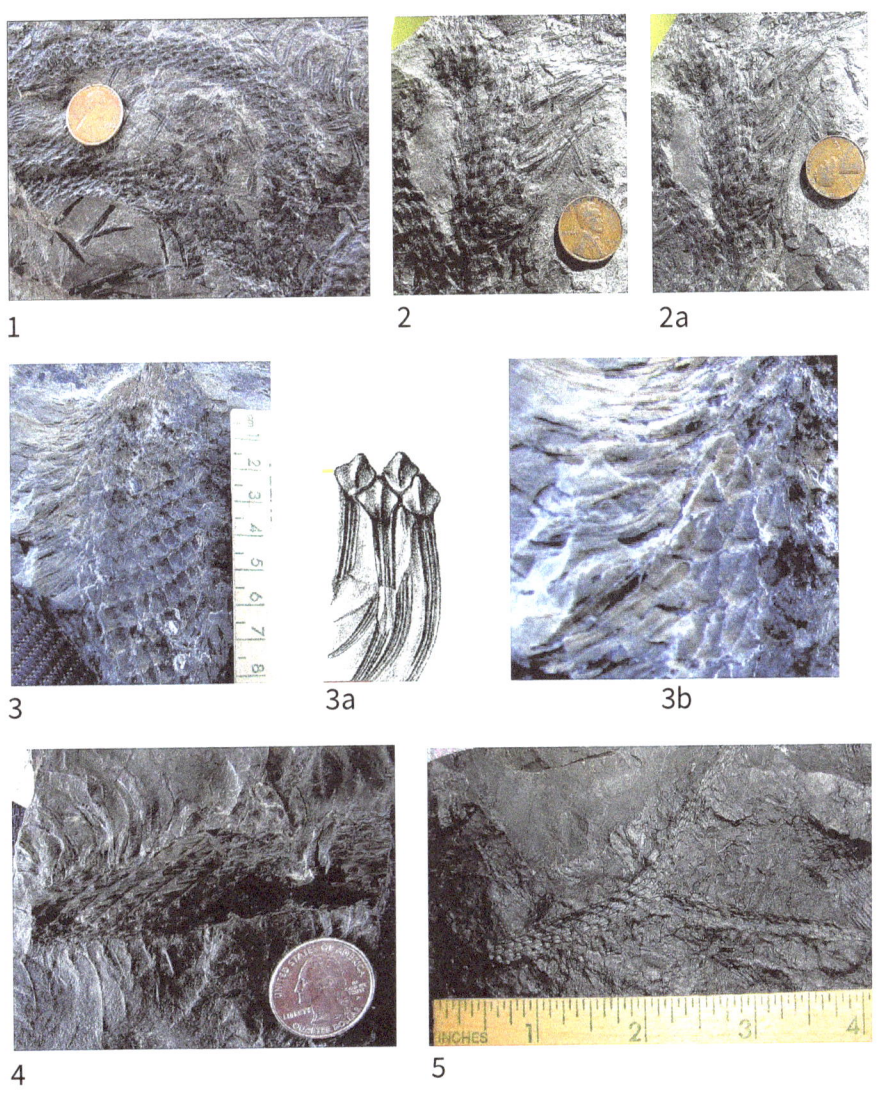

Plate I Lepidodendron Branches

Plate I. *Lepidodendron* foliage. 1. *Lepidodendron* cf. *wortheni* in sandy shale. Collected from a outcrop of the Blair coal seam in the Wise Formation along U.S. Route 58 Alternate, approximately 0.5 miles east of Appalachia High School, Appalachia, Wise County, Virginia.

2. *Lepidophylloides* sp. in siltstone. Collected from the roof strata of a mine in the Splashdam coal seam, located along Abners Fork (State Rt. 670) southeast off State Route 645 near Hurley in Buchanan County, Virginia.

3, 3a. *Lepidophylloides* in shale. 4. *Lepidophylloides* in shale. Collected from the roof strata of a mine in the Parsons coal seam located along Mud Lick Creek, 2.4 miles north of Roda in Wise County, Virginia.

Plate II. *Lepidodendron* foliage. 1, 1b. *Lepidodendron* twigs with attached *Lepidophylloides* in shale. 1a. Modern club moss branch for comparison. Collected from the roof strata of a mine in the Splashdam coal seam located along Abners Fork (State Route 670), southeast off State Route 645 near Hurley in Buchanan County, Virginia.

2. *Lepidostrobophyllum lancifolius* Lesquereux, 1870. 3. *Lepidostrobophyllum lanceolatus* Lindley and Hutton, 1831. Collected from the Kennedy coal seam in an outcrop of a road cut located 0.1 mile east of the junction of U.S. Route 58 Alternate and Boatright Hollow Road in Coeburn, Wise County, Virginia.

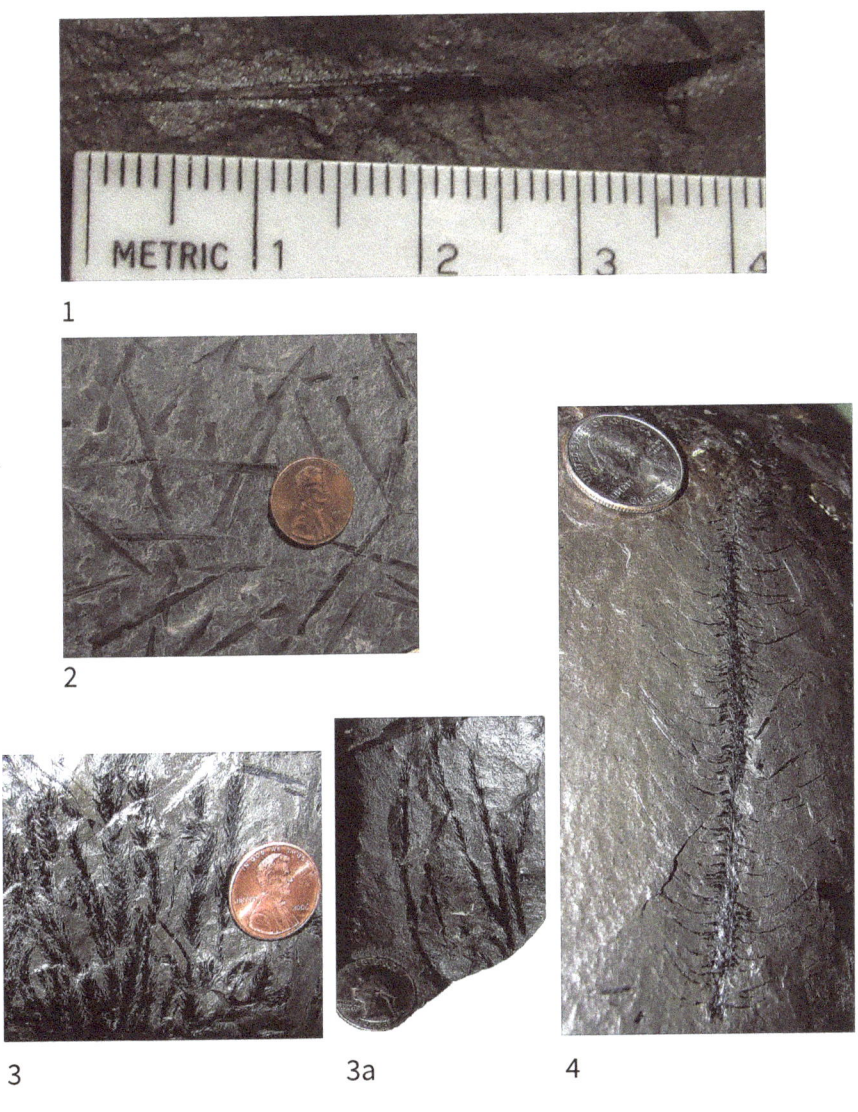

Plate I Lepidodendron Foliage

Plate II Lepidodendron Foliage

CHAPTER 2
ARBORESCENT LYCOPODS (CLUB MOSSES)

Sigillaria

Another club moss tree, *Sigillaria* (figure 6), along with its relative *Lepidodendron*, were among the most common and most widespread floras of Europe and North America. Both club mosses belong to the lycopod family. These trees dominated the Carboniferous up to the Middle–Late Pennsylvanian boundary. Progressively smaller forms existed through the Mesozoic era with the last surviving member of the group considered to be the modern quillwort (*Isoetes*).

Sigillaria and *Lepidodendron* were differentiated by the pattern and morphology of the leaf cushions and scars. The circular scars of *Sigillaria* were arranged in vertical columns in contrast to *Lepidodendron*, where the scars were spirally disbursed along the stems.

The leaves of *Sigillaria* were long and grass-like, forming circular scars or cushions as they were shed. The scars were arranged in vertical columns. Many species are identified on the basis of the shapes of the scars and the patterns of the scars. The impression of the bark of this plant is identified by broad linear ridges that are much wider than those of *Calamites*, and there is no segmentation. The ridges vary in design from plain to ornamental, with tiny circular depressions resembling bullseyes. Its leaves and roots are very similar to *Lepidodendron*, but it lacks the scale pattern on the trunk.

Like *Lepidodendron*, it grew to about 100 feet in height. This is truly remarkable considering the trunk of the tree consisted mostly of a spongy, weak tissue encased in a thin layer of a vesicular skin or woody bark.

Figure 6. Reconstruction of *Sigillaria* both branching and non-branching forms. The red objects represent reproductive organs. Modified after Gillespie et al., 1978.

Plate I. *Sigillaria.* 1. *Sigillaria rugosa* Brongn. 1a. Enlarged view showing the detail of the outer surface morphology. Specimen preserved in shale. Collected from a coal mine in the Lowsplint coal seam, 2.4 miles north of Stonega on Stonega Road (State Route 78) in Wise County, Virginia.

2. *Sigillaria, Mesolobus depressus* Stevens. 3. *Sigillaria* sp. 4. *Sigillaria mammilaris* showing the original outer surface (bark) and the underlying traces of the vascular bundles (parichnos) revealed as a result of decortication. These are the sites of foliage attachment (leaf scars) which were retained as the plant grew and dropped its leaves. Figures 2 and 3 are preserved in carbonaceous shale and figure 4 in dark gray shale. Collected from a coal mine in the Parson coal seam along Mud Lick Creek, 2.4 miles northeast of Roda in Wise County, Virginia.

5. *Asolanus comptotaenia* Wood. Possibly a decorticated *Sigillaria brardii*. Collected from the roof strata of a coal mine in the Splashdam coal seam located off State Route 610 near Conaway in Buchanan County, Virginia.

Plate II. 1. *Sigillaria* sp. preserved in fine grained sandstone. Collected from the seatrock immediately below the Hagy coal seam at an underground coal mine located along Grant Branch Road off State Route 619 (Lee Master Drive), southwest of Vansant in Buchanan County, Virginia.

2. *Sigillaria boblayi.* 2a. *Sigillaria elegans.* This is the reverse side of *Sigillaria* 2. These are examples of the subgenus *Eusigillaria*, group Favularia. The specimen, preserved in a dark gray shale, was collected from the roof strata of a coal mine in the Pocahontas No. 3 coal seam along Dog Fork Creek, 4 miles northeast of Cucumber in McDowell

County, Virginia, near the West Virginia state line.

Plate III. 1, 1a. *Sigillaria* sp. preserved in medium grained sandstone. Collected above the Aily coal seam, approximately 50 feet right off the westbound lane of U.S. Route 58 Alternate, 1.5 miles from Coeburn in Wise County, Virginia.

Plate I Sigillaria

1

2

2a

Plate II Sigillaria

1

1a

Plate III Sigillaria

Another variety or genus of the arborescent lycopods is the *Lepidophloios* (Gillespie, 1978). This was an erect, moss-like, evergreen plant resembling a tree in properties, growth, and structure, with spiraling leaf-cushions and appearance of the genus *Lycopodium*, like the club moss or ground pine (see *Lepidodendron* and *Sigillaria*). Shown in Plate I, *Lepidophloios*, are different aspects of an exceptionally well-preserved trunk of *Lepidophloios laricinus* (Dilcher, 2005). The fossil measures 9 inches in diameter by 24 inches in height. This specimen was collected by a coal miner, Samuel ("Sam") Knott, who lives in Bluefield, West Virginia. In 2002, it took both him and a friend to haul the fossil from a very remote area of a mine to the mine's shaft. From there it was hoisted 800 feet up the shaft to the surface. The mine is in the Pocahontas No. 3 coal seam near Pineville in Wyoming County, West Virginia.

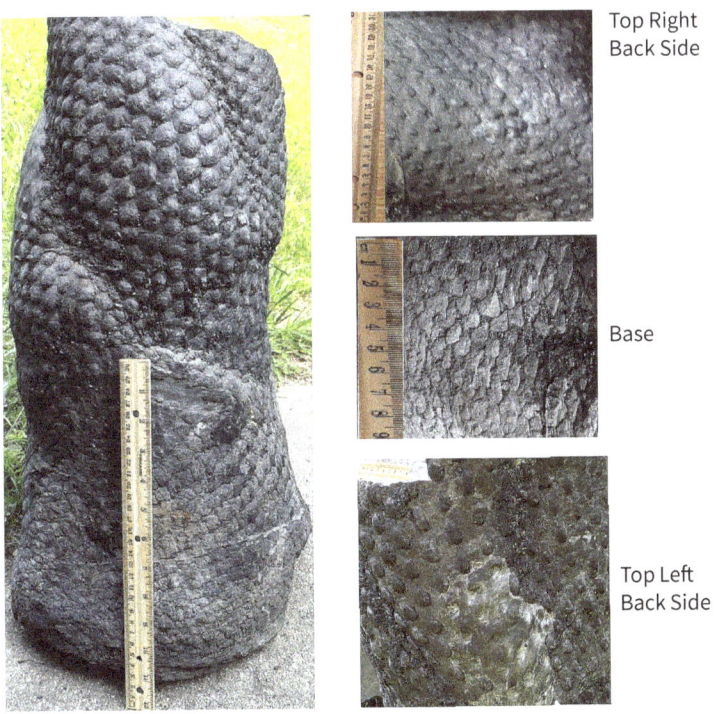

Plate I. Lepidophloios.

CHAPTER 3

CORDAITES: EARLY GYMNOSPERMS

Ancient Mangrove-Like Plant

Cordaites were trees that reproduced from seeds and spores borne by cone-like structures considered by some to be an "early conifer" or gymnosperm (see figure 7). They first appeared in the Upper Mississippian and then disappeared after the Triassic period. There are no extant descendants of *Cordaites*. Initially the name *Cordaites* was applied only to the narrow, strap-like compression leaf remains, but it has come to be applied to the entire plant. It is believed that one variety of the plant lived on dry land, growing to heights of up to 98 feet, while its shrub-like counterpart lived in marine-to-brackish water conditions on stilt-like root systems, much like the modern mangrove (see figure 7). Specimens of *Cordaites* have been found at only three stratigraphic horizons in the study area.

Figure 7. Reconstruction of the two varieties of *Cordaites*, including a mangrove-type (right). From Gillespie et al., 1978.

Plate I. *Cordaites*. 1. *Cordaites* stem in transition between the stages of decortication. *Artisia horizontalis* with pronounced longitudinal ribs (pith cast) and *Artisia transversa* with wrinkled surface. 1a, 1b. Enlarged views of specimen. 1c. *Artisia horizontalis* cast. 1d. *Artisia horizontalis* mold. 2. *Artisia transversa*. Note the thin surface layer of bark which has been carbonized. Collected from the Norton Formation immediately above an unnamed coal seam on U.S. Route 58 Alternate West along the railroad tracks in Appalachia, Wise County, Virginia.

Plate II. *Cordaites* foliage. 1. *Cordaites borassifolius* preserved carbonaceous in shale. 1a. Enlarged view of specimen 1 revealing the detail of the outer surface morphology. 2. Several *Cordaites borassifolius* superimposed on each other in shale. 2a. Enlarged view of 2 showing detail of morphology. Collected from the Upper Banner coal seam near Bucu in Dickenson County, Virginia.

3. *Cordaites (Noeggerathiopsis) hislopi* preserved in a black carbonaceous shale layer between coal and canal coal approximately 28 inches below the Norton coal seam rider. Collected from an outcrop at the junction of State Route 610 and Thackers Branch Road West near Norton in Wise County, Virginia.

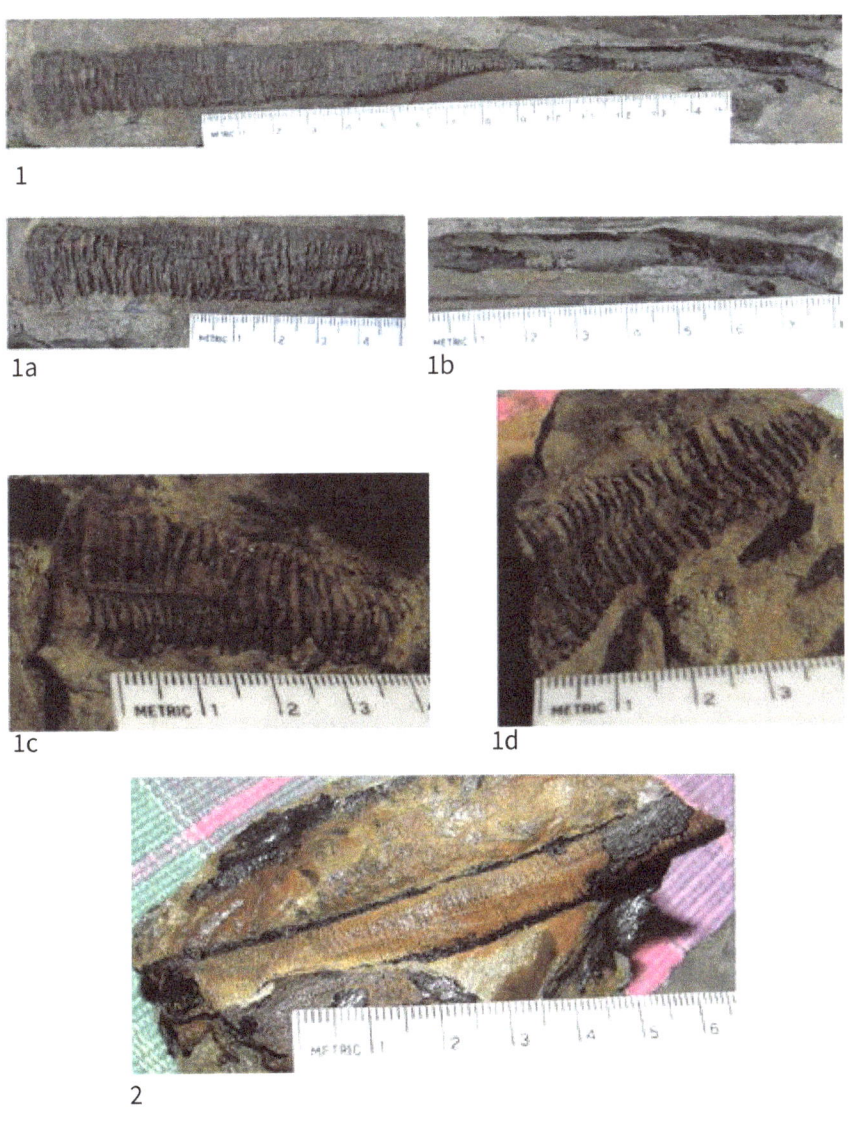

Plate I Cordaites Branches

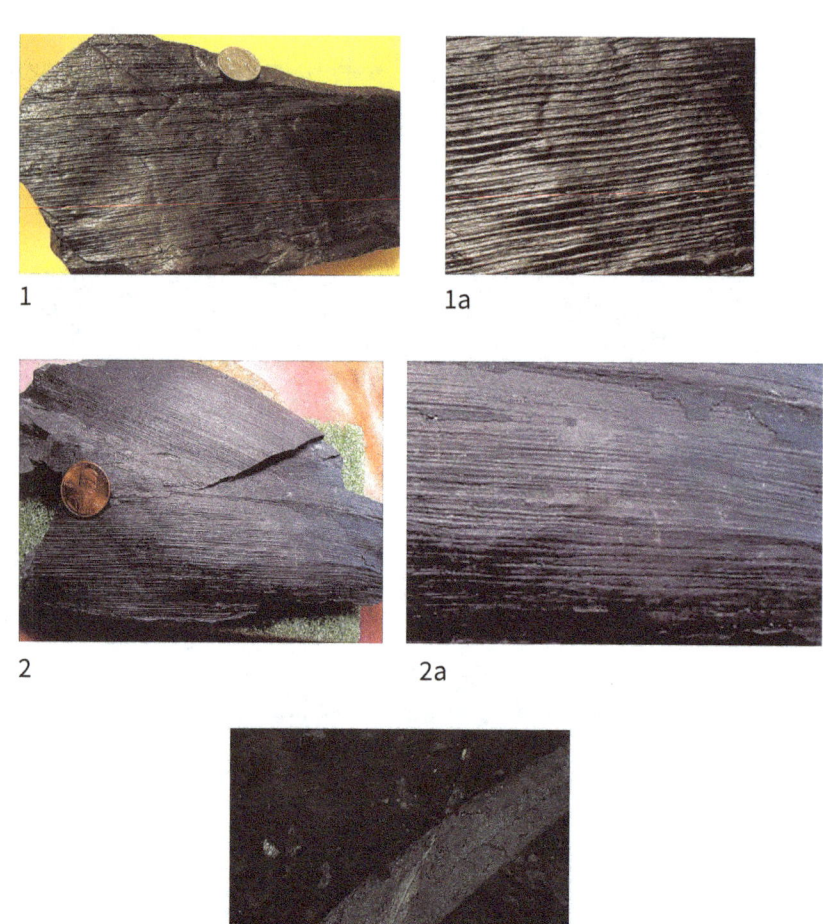

Plate I Cordaites Foliage

Plate I. *Cordaites* reproductive organs. 1, 1a. Mold and cast of *Cordaianthus* sp. preserved in silty shale. These are the reproductive organs. The single axis or shoot, sometimes referred to as "cones," bear the spore (male fructification) and the seeds or buds (female fructification). Collected from an outcrop of the Glamorgan coal bed in a road cut along Breaks Park Road, 5.7 miles northeast of Haysi in Buchanan County, Virginia.

1   1a

Plate I Cordaites Reproductive Organs

CHAPTER 4
KETTLEBOTTOMS

Ancient Tree Trunks

Kettlebottoms are the fossilized trunks of standing trees (see figure 8). They are characterized by being nearly circular in cross-section and often flare out near the base, giving the resemblance of a pot belly stove pipe (see figure 9). In this way, these fossils are often referred to as "stove pipes" by coal miners. Most have the bark preserved as a thin coal rind, which is usually slickensided, or highly polished, where clays surrounded the cast, compacted and lithified into shale. Some of the ancient tree trunks, including *Calamites*, were filled in by sand-sized particles of sediment.

Figure 8. Casts of two nearly complete lycopod tree trunks (arrows) that are approximately 15 and 19 feet tall. Note how the plant to the left appears to have been filled in by the siltstone beds about one-third of the distance from the tree base. They are in the high-wall (outcrop) of an underground mine developed in the Parsons coal seam, which is located in Pine Branch Hollow near Appalachia in Wise County, Virginia. A kettlebottom collected from the mine is shown in Plate I, Kettlebottoms, image 2.

Figure 9. Kettlebottom preserved in shale directly on the Clintwood coal seam at a surface mine in southwest Virginia. Note the contorted bedding that formed slickensided surfaces during differential compaction of the sediment as the tree was being buried. The carbonized bark is visible as a thin black layer of coal between the cast and the surrounding strata (arrow).

A very unusual preservation of two kettlebottoms is shown in Plate I, kettlebottoms 2 and 3. Even more noteworthy is the unique preservation of the plant fibers that made up the bark of the tree, which is shown in Plate I, kettlebottom 3a. It should be noted that the red color of kettlebottom 3 is natural, not painted on. The geologic sequence in the development of fossilization of tree trunks, as well as that of *Calamites*, is illustrated in figure 10.

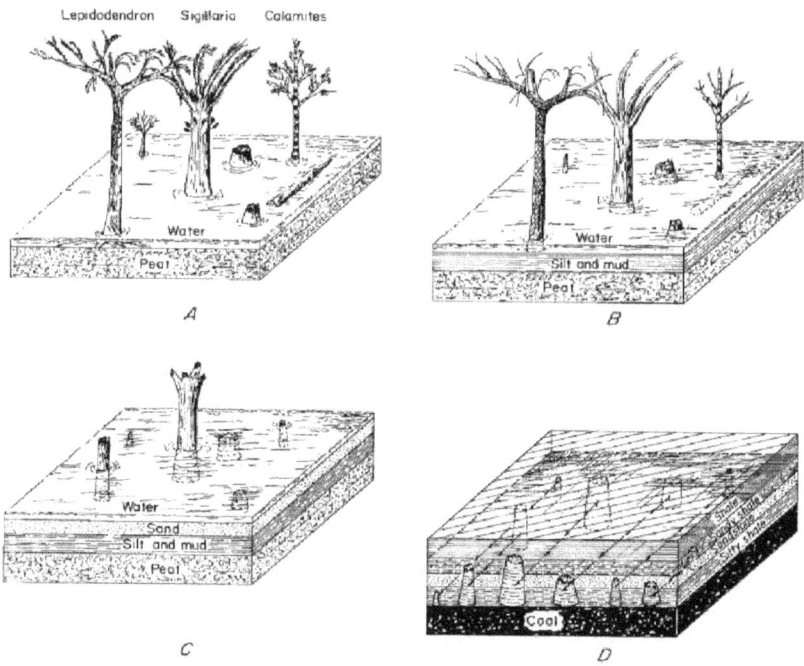

Figure 10. A generalized sequence of events leading up to formation of kettlebottoms. A. Plants are growing in a typical swamp-forest. B. Flooding of the swamp by sediments overcomes peat formation, choking and killing the plants. C. Dead plants decay to become hollow and eventually break off at water level. D. Continued sedimentation fills in the stumps, followed by compaction and cementation of the sedimentary particles and peat, which results in the formation of coal, rock (sandstone or shale), and kettlebottoms. (From U.S. Bureau of Mines, Report of Investigation 8785, 1983, p. 4.)

Plate I. Kettlebottoms. 1. Kettlebottom preserved in gray, slickensided shale from a mine in the Taggart coal seam near Imboden in Wise County, Virginia.

2. Kettlebottom with stigmaria attached, preserved in sandstone. Collected from the Upper Banner coal seam, 3.2 miles north of Dante, Virginia, and left off State Route 63 North. The specimen is located at the Yale Peabody Museum of Natural History in New Haven, Connecticut.

3. Kettlebottom of *Sigillaria* preserved in sandstone that was exposed to very high heat during burial, which resulted in its reddish appearance. Collected from the Upper Banner coal seam, 3.2 miles north of Dante, Virginia, and left off State Route 63 North. The specimen is located at the Yale Peabody Museum of Natural History in New Haven, Connecticut. 3a. Enlarged view of the bark of the kettlebottom shown in image 3, detailing the fibrous structure just beneath the layer that was the bark of the tree.

4. Kettlebottom preserved in gray, slickensided shale from a mine in the Parsons coal seam located in Pine Branch Hollow, near Appalachia in Wise County, Virginia.

5. A portion of the trunk of *Psaronius schopfiim* preserved in gray, coarse-grained sandstone. Collected from the seatrock immediately below the Hagy coal seam in an underground mine located along Grant Branch Road off State Route 619 (Lee Master Drive), southwest of Vansant in Buchanan County, Virginia.

Plate I Kettlebottoms

CHAPTER 5

STIGMARIA

Ancient Root Systems

Stigmaria are the fossilized root stocks of *Lepidodendron* and *Sigillaria* that originated from the lower trunk of the plant and penetrated the clays, silts, or sands. These sediments formed the paleosoils of the swamps and forests. *Stigmaria* are generally characterized by circular, pit-like scars from which rootlet hairs emerged (see figure 11a). The scars vary in size from a pencil point up to a pencil eraser in diameter. Often the rootlet hairs are represented by randomly arranged, narrow, linear groves that appear to wrap around the organ. It was through the tubes that the lycopod absorbed its nutrients from the marsh sediments, which lithified into shale, siltstone, and sandstone. This is known as the paleosoil or "seatrock." Examples of the fossil root systems are presented in *Stigmaria* plates I–III. A modern root system of a typical tree and its internal anatomy are shown in figure 11b. A rare preservation of the "pith" cast of a *Lepidodendron stigmaria* is presented in plate II, image 1b, which appears as a second layer or "growth ring" with morphology similar to the surface layer.

A   B

Figure 11. Modern tree root system including rootlets that penetrated into the soil and were exposed to the surface by erosion (A). Graphic rendition of the generalized internal structure of the root showing the inner core, referred to as the "pith" (B).

Plate I. *Stigmaria*. 1. *Stigmaria ficoides*. Note the tube-like rootlets that extend laterally and vertically in the gray shale from the root. The shale was originally clay (mud) in which the root of the tree grew. This is known as the paleosoil, which, when solidified, is referred to as the "seat rock" immediately below the coal seam. Collected from a mine in the Splashdam coal seam located along Abners Fork Road (State Rt. 670), southeast off State Route 645 near Hurley in Buchanan County, Virginia.

2. *Stigmaria ficoides*. Note the numerous nearly circular scars, which were sites of rootlet attachments to the bark of the root system, resulting in a stipple-type morphology. 3, 4. *Stigmaria eveni* Lesquereux, 1866A. Collected from a mine in the Parsons coal seam, 2.4 miles north of Roda in Wise County, Virginia, on Mud Lick Creek.

5. *Stigmaria ficoides* preserved in a silty gray shale. 5a. Enlarged view showing detail of rootlet scars configuration. Collected from the seatrock immediately below the Hagy coal seam at an underground coal mine located along Grant Branch Road off State Route 619 (Lee Master Drive), southwest of Vansant in Buchanan County, Virginia.

Plate II. 1, 1a, 1b. *Stigmaria ficoides* preserved in a fine-grained, carbonaceous sandstone. The cast of a *Lepidodendron* tree root. Image 1 is a view of the internal pith cast. Collected from a mine in the Lower Banner coal seam on Neece Creek, south of Nora in Dickenson County, Virginia.

2. Root-like system. 2a. End view showing layers of carbonization, which may be a form of growth ring. Specimens preserved in silty shale. Collected from the seatrock of the Blair coal seam outcrop along U.S. Route 58 Alternate, approximately 1 mile east of Appalachia in Wise County, Virginia.

3. Root Cast of *pteridosperm* (seed fern) in silty shale. 4. Root Cast of *pteridosperm* associated with *Alethopteris decurrens* fern fronds. Collected from a mine in the Parsons coal seam, 2.4 miles north of Roda in Wise County, Virginia, on Mud Lick Creek.

Plate III. 1. *Stigmaria ficoides* preserved in a fine-grained sandstone. 1a, 1b. Cross-sectional views showing the internal pith cast or core of the root. Collected from the Clintwood coal seam along I-23 North in Norton, Wise County, Virginia.

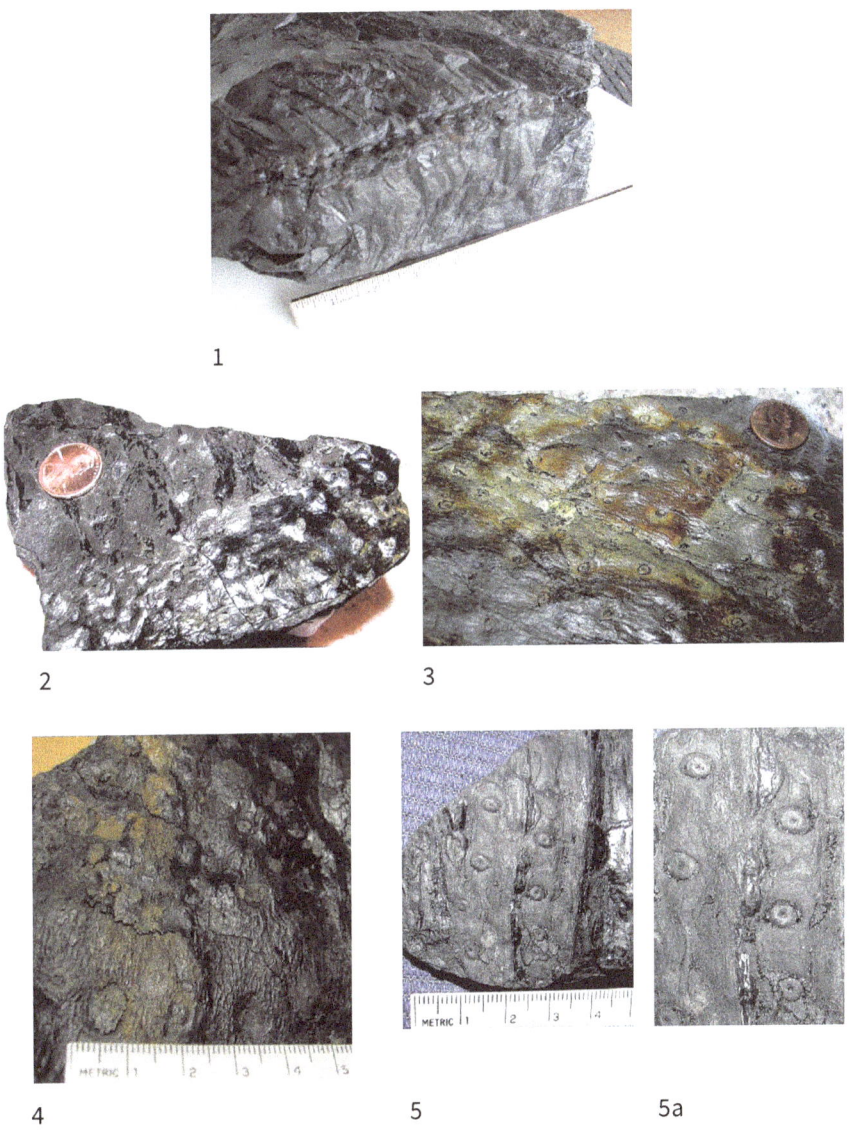

Plate I. Stigmaria

Plate II. Stigmaria

1

1a   1b

Plate III. Stigmaria

CHAPTER 6
CALAMITES

Ancient Relative of the "Horsetail"

Calamites (see figure 12a) are the extinct ancestors of the modern sphenophytes (horsetails). The basic anatomy of the calamites is shown in figure 12b. A few examples of sphenophytes are shown in figure 13. Unlike its living relative, calamites grew to the size of small trees as much as six inches in diameter. Its main stalk or trunk has a bamboo-like appearance characterized by being segmented (jointed) at intervals with finely spaced grooves oriented parallel to the long axis of the plant, as illustrated in figure 14.

The rib patterns of the preserved pith casts where they meet at the nodal line are used to group the genera and subgenera of *Calamites*. Along the joint, one can see radial scars where smaller branches split off of the main stem (see *Calamites* plates I and II). Specimens in which the ribs alternate are true *Calamites*; ribs that have some alternating with others passing through the node are assigned to the subgenus *Mesocalamites*. The leaves are distributed in a circular (or whorled) pattern around the stems at evenly spaced intervals, somewhat resembling a pinwheel. Specimens with leaves that are linear, lanceolate, or spathulate are referred to as *Annularia*. Their bases form a collar around the stem but could be absent from some fossils; they have leaves in whorls of 5–32 per node. *Asterophyllites* have leaves that are longer and narrower than those from *Annularia*. These leaves are not united at the bases and have whorls of 4 to 40 per node; they arch steeply upward from the stem.

A genus that is similar to *Calamites* but differs in the foliage and strobili is called *Archaeocalamites* (see figure 15).

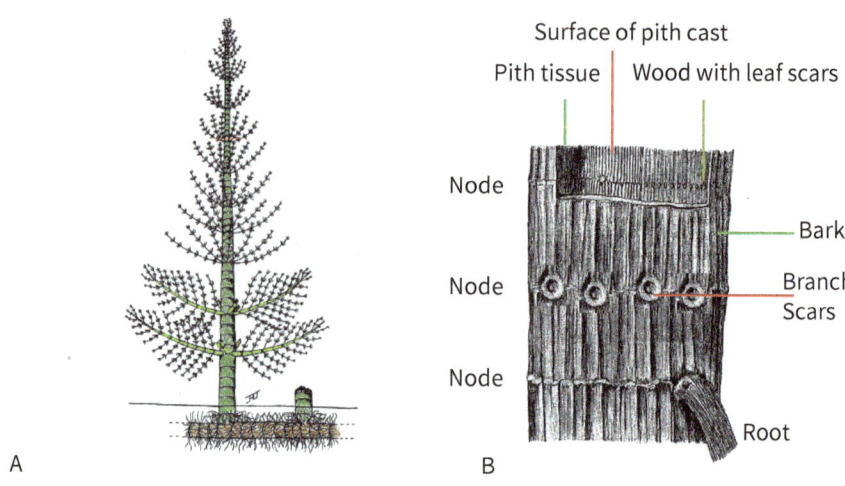

Figure 12. A. Reconstruction of Calamites. Modified after Gillespie et al.1978. B. Artist's reconstruction of a portion of the main trunk of Calamites (Calamitina) showing the basic internal and external features preserved in rock as a fossil. Modified from Seward, 1898, vol. I, p. 316, fig. 77.

Figure 13. Examples of modern relatives of Calamites.

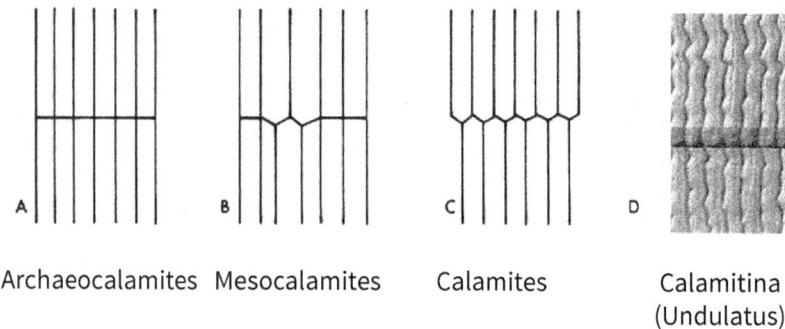

Archaeocalamites   Mesocalamites   Calamites   Calamitina
                                                                 (Undulatus)

Figure 14. Calamites stem characteristics A through C are used to identify the general genera. D is the subgenera Calamitina, which is associated with t he species Undulatus. Modified after Gillespie et al., 1978.

Plate I. *Calamites*. 1. *Calamites cisti* preserved in carbonaceous shale. 1a. Enlarged view showing detail of ribs and node. Collected from the roof strata in a mine in the Lowsplint coal seam located near Stonega in Wise County, Virginia.

2. *Calamites* sp. Impression in shale collected from the roof strata in the Pocahontas No. 3 coal seam on Dog Fork, 4 miles northeast of Cucumber in McDowell County, West Virginia.

3. *Calamites (calamitina)*. The bark and impression of the wood in a medium-dark brown shale. 3a. Enlarged view showing detail of the *calamitia* (i.e., type of branch scar pattern) along the node. Collected from the roof strata of a mine in the Splashdam coal seam located along Smith Branch (State Route 701), 0.7 miles north of Slate Creek (State Route 83), 8 miles east of Grundy in Buchanan County, Virginia.

4. *Calamites ramosus* showing three sites of branch attachments or branch scars. 4a. Enlarged view of one of the branch scars. Collected from the roof strata of a mine in the Parsons coal seam, 2.4 miles north of Roda in Wise County, Virginia, on Mud Lick Creek.

5. *Calamites cisti* preserved in sandstone associated with an unnamed coal seam found in the Norton Formation. Collected from an outcrop along a railroad right of way parallel to the westbound lane of U.S. Route 58 Alternate in Appalachia, Wise County, Virginia.

Plate II. *Calamites*. 1. *Calamites sp.* preserved in shale. 1a. Enlarged view showing the detail of ribs and node. Collected from the roof strata of a mine in the Jawbone coal seam located east of South Clinchfield in Russell County, Virginia.

2, 2a, 2b. *Calamites undulates* preserved in shale. The arrow in image 2 points to the site of one of four branch scars. The arrow in image 2b points to the site of one of the two tiny branch scars at one node. Image 2b is a fragment of a specimen similar to 2a, showing detail of the rib morphology which the species name implies. Collected from the roof strata of a mine in the Parsons coal seam, 2.4 miles north of Roda in Wise County, Virginia, on Mud Lick Creek.

3, 3a. *Mesocalamites* sp. preserved in shale. Found in the upright growth position in the strat just above the Kennedy coal seam outcrop, located near the southbound lane of U.S. Route 58 Alternate, 1.5 miles east of Coeburn in Wise County, Virginia.

4. *Calamites undulates* preserved in shale. Collected from the strata above the Blair coal seam outcrop, located along the northbound lane of I-23 in Norton, Wise County, Virginia.

5. *Calamites carinatus*, which is a *Eucalamite (Diplocalamites)*. 5a. Enlarged view of the characteristic morphology of the terminus of the ribs at a node. Preserved in fissile shale. 5a. Enlarged view showing the detail of ribs and node. Collected from the roof strata of a mine in the Lower Banner coal seam, north of South Clinchfield in Russell County, Virginia.

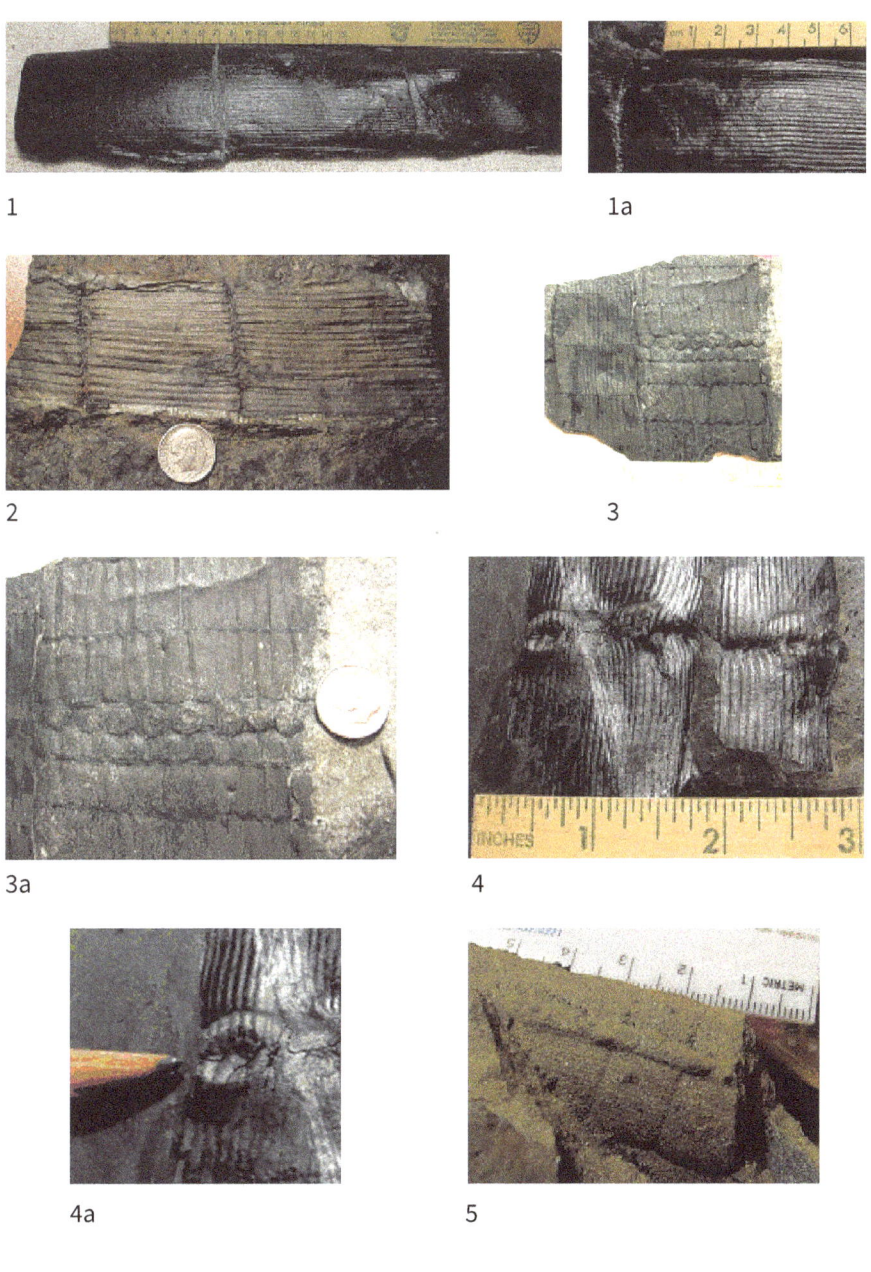

Plate I. Calamites

Plate II. Calamites

Plate III. *Calamites*. 1. *Calamites (calamitina)* preserved in a dark brownish-red iron stone concretion. Collected from the Blair coal seam horizon at the junction of I-23 and U.S. Route 58 Alternate in Wise County, Virginia.

2. *Calamites undulates* preserved in a medium gray siltstone. Collected from the Blair coal seam outcrop, approximately 500 feet northeast of the junction of U.S. Route 58 Alternate West and I-23 near Norton in Wise County, Virginia.

3. *Calamites (Eucalamites) cruciatus* Sternberg preserved in dark gray siltstone. 4. *Calamites cisti* Brongniart preserved in a concretion. Collected from the Clintwood coal seam outcrop, located along I-23 South at the Wise Shopping Center in Wise, Wise County, Virginia.

Plate IV. *Calamites* (new species?). 1, 1a. *Calamites* sp.? This specimen appears to be a variant of the *mesocalamites* in that there is one prominent rib wedged between sets of ribs of much smaller size rather than ones which are equal in size as the rib indicated by red arrow. Also, ribs in contact with the former should not be continuous across the nodal zone. Therefore, this may be a new species. Collected from a montmorillonite clay bed just above the Aily coal seam horizon, 1.5 miles west of Coeburn in Wise County, Virginia, along U.S. Route 58 Alternate.

Plate V. *Calamites*. 1. *Calamites undulatus* cf. preserved in a silty yellowish-brown shale. Note the sharp contrast in rib thickness and spacing on the left (red arrow) vs. the right side (blue arrow) of the specimen. 2. *Calalmites ramifer* Stur, 1875, preserved in medium yellowish-brown shale. There are no furrows; the ribs are perfectly flat and the bark paper thin. 3. *Calamites undulates* cf. displaying contrasting morphology of the outer layer (black arrow) and inner layer (green arrow) of plant tissue. 4. *Calalmites ramifer* Stur, 1875, with branch scar (arrow). 5. *Calamites* sp. with branch attached (arrow). Collected from the Phillips coal seam, located in a road cut 4.8 miles northwest of Inman in Wise County, Virginia, on Route 160. Note: The preservation of images 2, 4, and 5 were in the form of a very thin layer (like parchment paper) on the rock.

Plate VI. *Calamites*. 1. *Mesocalamites undulatus* cf. preserved in a gray shale. 2. *Calamites undulatus*. 2a. *Asterophyllites charaeformis* found on reverse side of image 2. 3. *Calamites undulatus*. Rather large specimen preserved in a medium gray shale at the same horizon as images 1 and 2. Collected from the Phillips coal seam, located in a road cut 4.8 miles northwest of Inman in Wise County, Virginia, on Route 160.

Plate III. Calamites

1

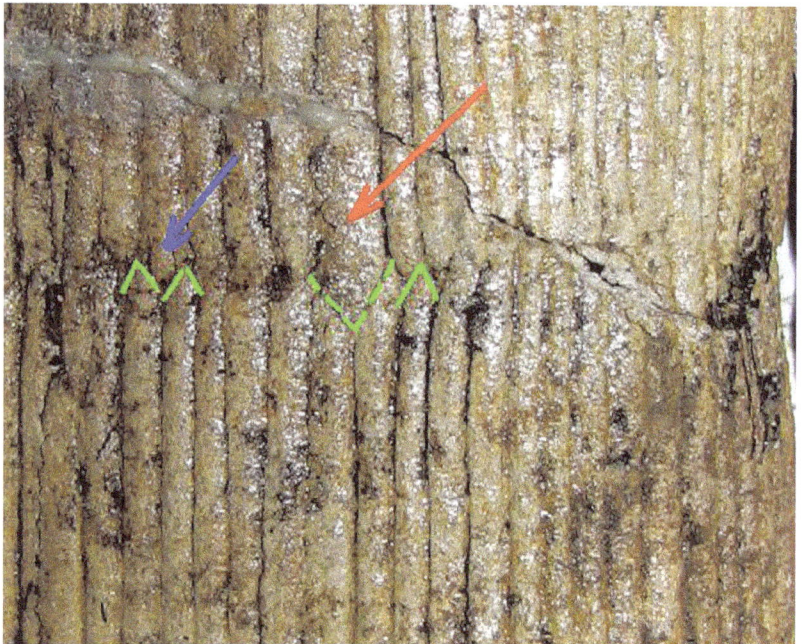

1a

Plate IV. Calamites (new species?).

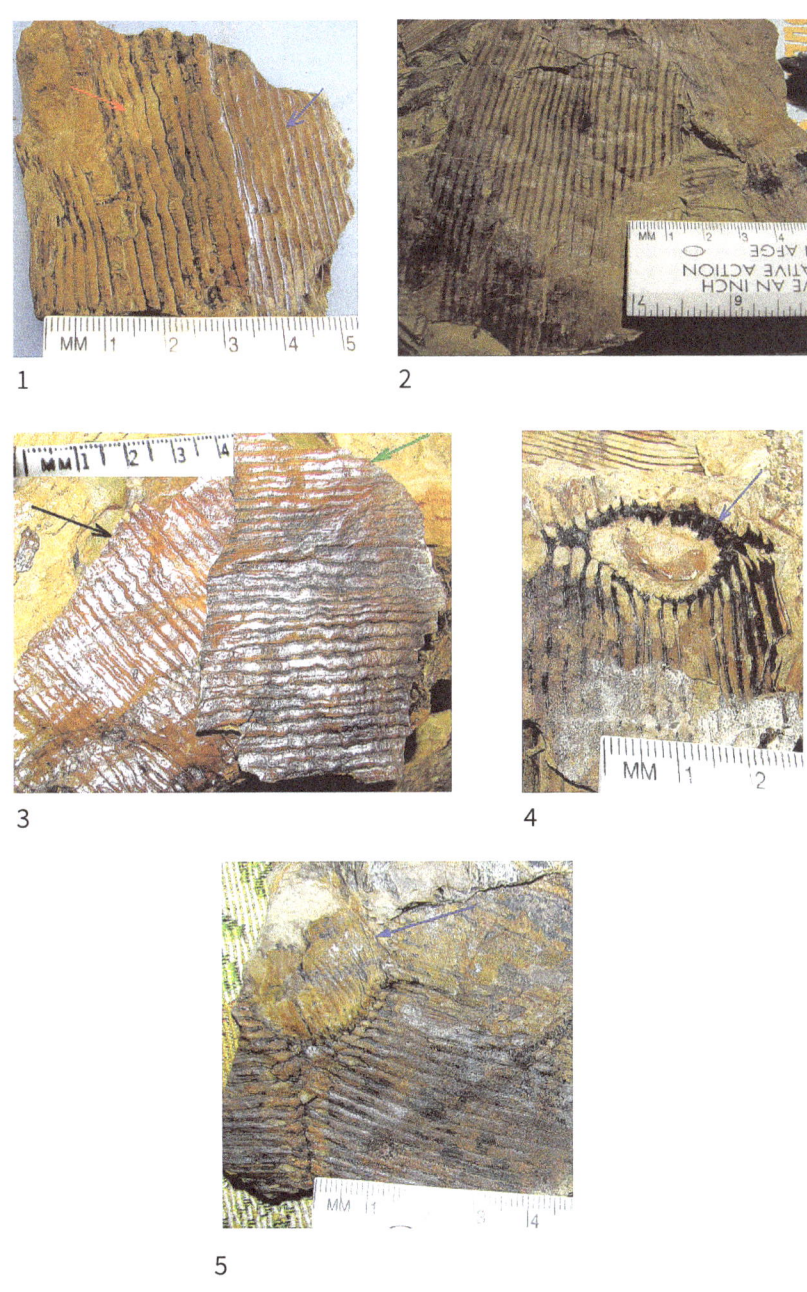

Plate V. Calamites.

1

2

2a

3

Plate VI. Calamites.

Plate VII. *Calamites*. 1. *Calamites suckowi* (base of the stem) preserved in a dark gray argillaceous siltstone. 2. *Calamites suckowi*, showing detailed morphology at the nodes, preserved in dark brown siltstone. Collected from the Norton coal seam outcrop, located approximately 2000 feet west of the junction of U.S. Route 23 Alternate and Route 83 near Pound in Wise County, Virginia.

3. *Calamites* sp. preserved in a dark gray siltstone. 4. *Calamites* sp. preserved in a medium gray siltstone. 4a. Enlarged view showing surface punctures caused by roots that had grown through it. Collected from an outcrop of the Norton coal seam, 0.6 miles west of the junction of State Routes 643 and 641 on Bold Camp Mountain, 2 miles south of Pound in Wise County, Virginia.

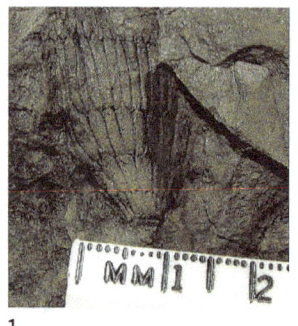

1

2

3

4a

5

Plate VII. Calamites

Plate I. *Calamites* foliage. 1, 1a. *Lobatannularia.* Collected from the shale immediately above the Kennedy coal seam outcrop, 1.5 miles from Coeburn in Wise County, Virginia, along U.S. Route 58 Alternate's eastbound lane.

2. *Annularia pseudostellata* in a siltstone from the Clintwood coal seam horizon along I-23 South behind the shopping center in Wise, Wise County, Virginia.

3 *Annularia asteris.* 4. *Annualaria radiata.* Both preserved in a clay shale directly above the Blair coal seam from an outcrop along I-23 North near Norton in Wise County, Virginia.

5. *Asterophyllites charaeformis* preserved in a clay shale. Collected from the roof strata of a mine in the Lowsplint coal seam, 2.4 miles north of Stonega on Stonega Road, State Route 78, in Wise County, Virginia.

Plate II. *Calamites* foliage. 1. *Asterophyllites longifolius* in shale. Collected from the roof strata of a mine in the Lowsplint coal seam, 2.4 miles north of Stonega on Stonega Road, State Route 78, in Wise County, Virginia.

2. *Asterophyllites longifolius.* 3. *Equisetites hemingwayi* Kidst. Collected from the roof strata of a mine in the Parsons coal seam along Mud Lick Creek, 2.4 miles northeast of Roda in Wise County, Virginia.

4. *Equisetites hemingwayi* Kidst. Collected from the shale immediately above the Kennedy coal seam outcrop, 1.5 miles east of Coeburn in Wise County, Virginia, along U.S. Route 58 Alternate.

Plate III. *Calamites* foliage. 1. *Annularia stellata* (Schlotheim) Wood. Collected from the roof strata of a mine in the Parsons coal seam along Mud Lick Creek, 2.4 miles northeast of Roda in Wise County, Virginia.

2, 3. *Asterophyllites equisetiformis* in shale. Collected from the roof strata of a mine in the Splashdam coal seam, located along Abners Fork Road (State Route 670), southeast off State Route 645 near Hurley in Buchanan County, Virginia.

4. *Asterophyllites grandis*. Collected from the shale immediately above the Kennedy coal seam outcrop, 1.5 miles east of Coeburn in Wise County, Virginia, along U.S. Route 58 Alternate.

Plate I. Calamites foliage.

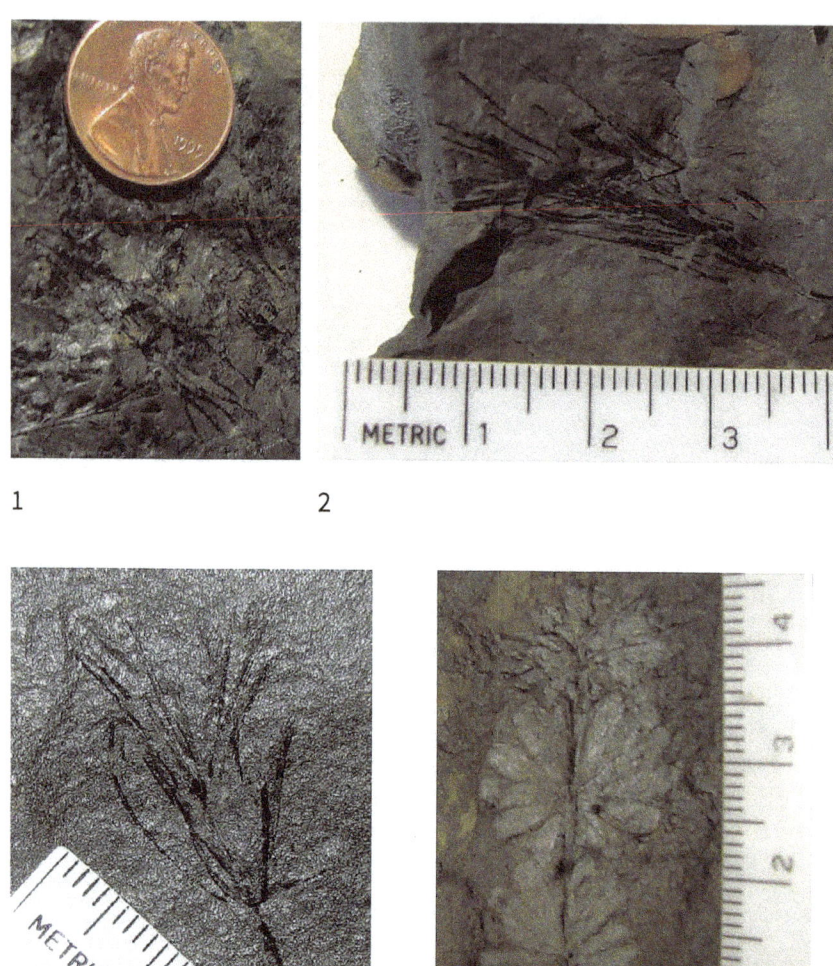

Plate II. Calamites foliage.

Plate III. Calamites foliage

Plate IV. *Calamites* foliage. 1. *Asterophyllites charaefomis* in a light gray silty shale. Collected from the Glamorgan coal seam outcrop in a road cut along Breaks Park Road, 5.7 miles northeast of Haysi in Buchanan County, Virginia.

2, 3. *Asterophyllites charaefomis* in a light yellowish-brown silty shale. Collected from the Phillips coal seam outcrop in a road cut 4.8 miles northwest of Inman in Wise County, Virginia, on Route 160.

4. *Annularia psuedostellata* in a grayish-white silty shale. Collected from the Phillips coal seam outcrop in a road cut 4.8 miles northwest of Inman in Wise County, Virginia, on Route 160.

5. *Equisetum telmatie* is a modern relative, shown for comparison with image 4. 6. *Asterophyllites charaeformis* preserved in a medium gray shale. Collected from the Phillips coal seam, located in a road cut 4.8 miles northwest of Inman in Wise County, Virginia, on Route 160.

Plate V. *Calamites* foliage. 1, 2. *Annularia radiata* in a light yellowish silty shale. Collected from the Blair coal seam outcrop along I-23 North near Pound in Wise County, Virginia.

Plate VI. *Calamites* foliage. 1. *Annularia radiata* in siltstone. Collected from the roof strata of a mine in the Splashdam coal seam, located along Smith Branch (State Route 701), 0.7 miles north off Slate Creek (State Route 83), 8 miles east of Grundy in Buchanan County, Virginia.

2, 3. *Annularia spicata* in a silty shale from above an unnamed coal seam, 5.0 miles northwest of Inman in Wise County, Virginia.

Plate IV. Calamites foliage

Plate V. Calamites foliage

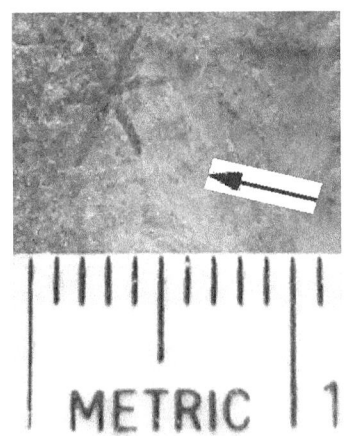

1

2

3

Plate VI. Calamities foliage

Plate I. *Calamites* cones. 1. *Bowmanites* Binney preserved in shale along with associated *Sphenophyllum* leaves, lower left-hand corner. 1a. Enlarged view of a cone from image 1. The specimens were collected from the strata immediately above the Kennedy coal seam in an outcrop along I-23 South, 1.5 miles east of Coeburn in Wise County, Virginia.

2. *Asterophylittes charaeformis* (Green) cone, associated with *Annularia radiata* (Blue). Preserved in a yellowish-brown shale.

3. *Palaeostachya*. 4. *Calamostachya*. Collected from the Phillips coal seam outcrop in a road cut 4.8 miles northwest of Inman in Wise County, Virginia, on Route 160.

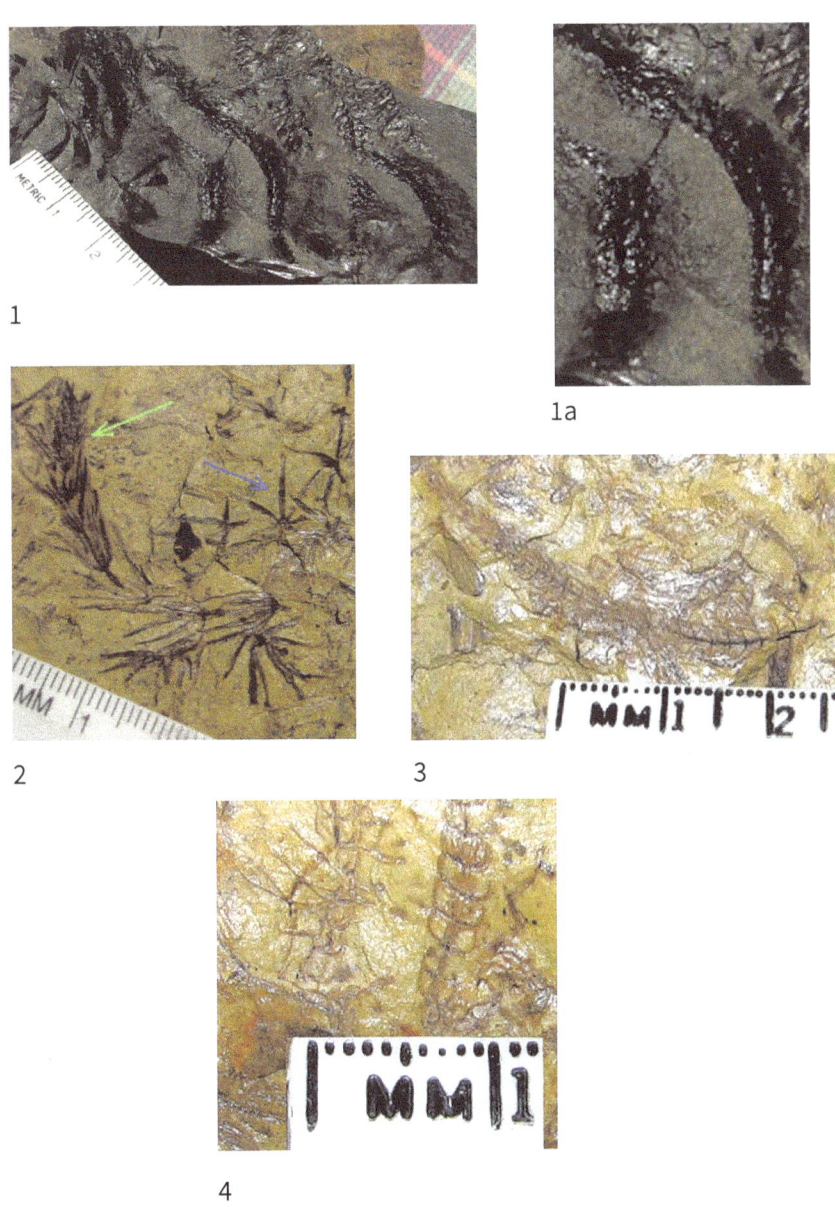

Plate I. Calamites reproductive cones.

Archaeocalamites is a Devonian and Lower Carboniferous plant that somewhat resembles *Calamites*. It differs in that the ribs and groves of the pith-cast, which can be slightly elevated to almost flat, exactly match up at the very slightly constricted nodes. The similarity between this plant and *Calamites* is in the internal structure. Contrasts in the characteristics of the leaves and cones seem to justify the placement of *Archaeocalamites* in a separate generic designation. Specimens of the foliage or reproductive organs were not found, but a small specimen of a stem was collected and is shown below, along with a seed pod belonging to another type of plant.

Figure 15. ArchaeoCalamites (reconstructed) with a seed pod preserved in a medium-grained sandstone. Collected from the Wise Formation immediately above an unnamed coal seam in an outcrop on U.S. Route 58 Alternate West along the railroad tracks in Appalachia, Wise County, Virginia.

CHAPTER 7

SPHENOPHYLLUM

Sphenophyllum are a genus of small, vein-like, and bramble-like land plants possessing characteristics that could be mistaken for the whorl from *Calamites* foliage, but typically smaller than *Annularia* or *Asterophyllites*. The stems were jointed and longitudinally ribbed. The foliage consisted of whorls of leaves that were triangular-shaped and rounded or forked at the apex. Most likely they resembled the modern ground covering (bed straw) plant *Galium* (see figures 16 and 17). Compare with the fossil plants shown in *Sphenophyllum* Plate I, image 4, and *Sphenophyllum* Plate II, image 1.

Figure 16. G. verum with needle-like leaves, 6 to a whorl. Compare this with Plate III Sphenophyllum, image 2, and see the resemblance.

Figure 17. G. aparine with rounded wedge-shaped leaves.

Plate I. *Sphenophyllum*. 1 *Sphenophyllum emarginatum*. 2.. *Sphenophyllum cuneifolium*. 3. *Sphenophyllum* cf. *myriophyllum*. 4. *Sphenophyllum cuneifolium*. 5. *Sphenophyllum* cf. *myriophyllum*. 6. *Sphenophyllum cuneifolium*. 7. *Sphenophyllum majus*. 8. Single leaf of *Sphenophyllum costae sterzel* showing webbing. 9. *Sphenophyllum emarginatum*. Collected from the shale immediately above the Kennedy coal seam outcrop, 1.5 miles east of Coeburn in Wise County, Virginia, along U.S. Route 58 Alternate.

Plate II. *Sphenophyllum*. 1. *Sphenophyllum* sp. *emarginatum* preserved in a yellowish-orange shale. Collected from an outcrop in a road cut along Breaks Park Road, 5.7 miles northeast of Haysi in Buchanan County, Virginia.

2. *Sphenophyllum cuneifolium* preserved in a yellowish-brown shale. Collected from strata above the first coal rider above the Blair coal seam in an outcrop along I-23 North in Norton, Wise County, Virginia.

3. *Sphenophyllum emarginatum* preserved in a light yellowish shale. Collected from strata above the Phillips coal seam, located in a road cut 4.8 miles northwest of Inman in Wise County, Virginia, on Route 160.

4. Enlarged view of leaves from *Sphenophyllum majus* showing webbing. Also see Plate I, *Sphenophyllum*, image 7. Collected from the shale immediately above the Kennedy coal seam outcrop, 1.5 miles east of Coeburn in Wise County, Virginia, along U.S. Route 58 Alternate.

Plate III. *Sphenophyllum*. 1, 1a. *Sphenophyllum cunefolium* f. *saxifragifolia* (Tenchov) preserved in a very fine-grained sandstone. Collected from strata above the Low Splint coal seam in an outcrop located 10.2 miles north of Norton in Wise County, Virginia, along State Route 620.

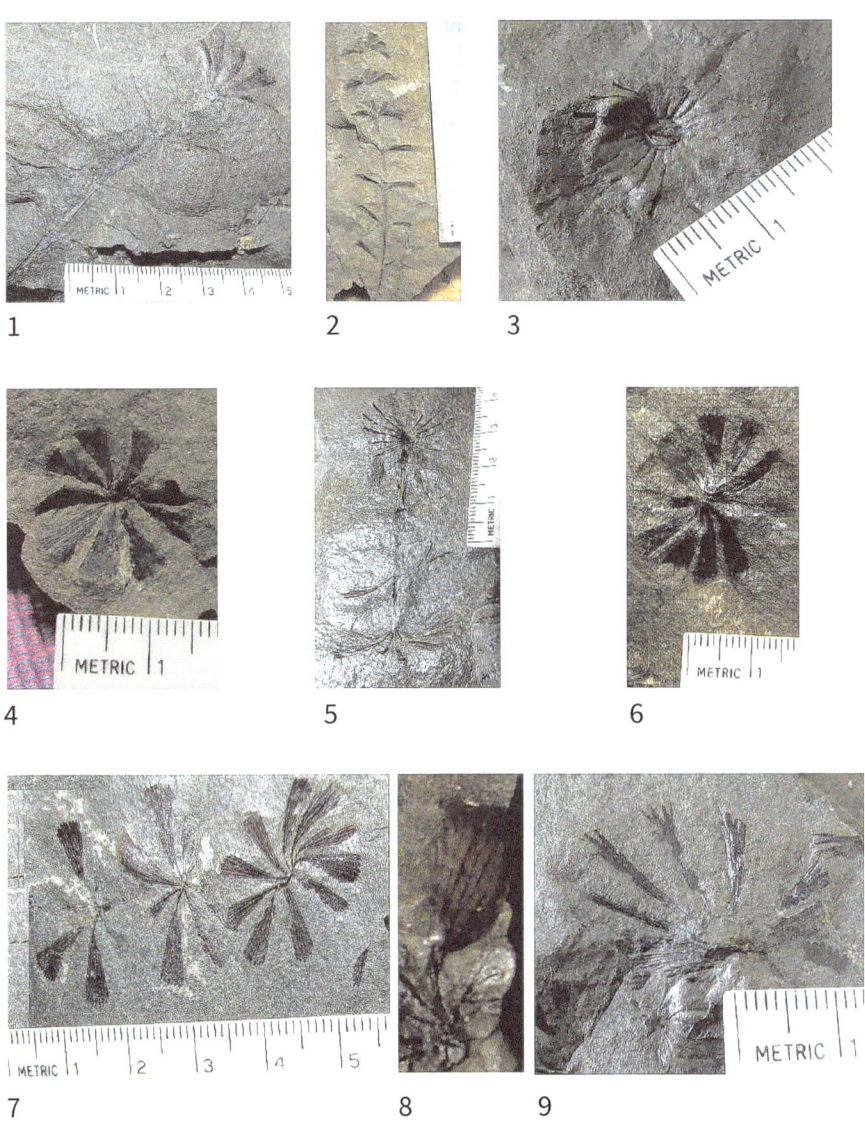

Plate I. Sphenophyllum.

1

2

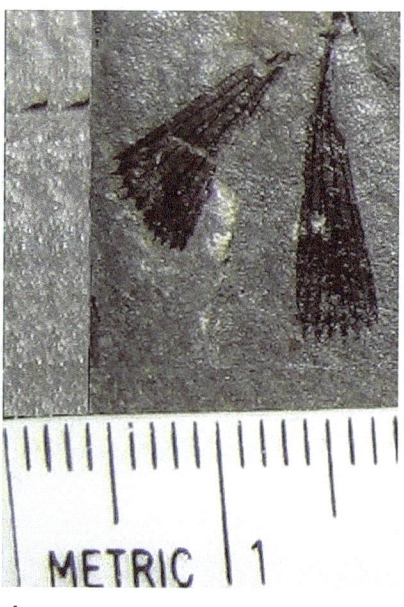

3

4

Plate II. Sphenophyllum.

1

2

Plate III. Sphenophyllum.

CHAPTER 8
FERNS

Fern Morphology - General

Classification and naming of ferns are based on a standard, specialized terminology that is used in even the most elementary field guides to fossil ferns. Figure 18 illustrates the basic terms applied to each of the components found in modern ferns; ancient ferns have the same terminology applied to them. Incomplete or fragmented fern-like compound leaves are assigned to the basic form-genera determined by using the general morphology of pinnules, venation, and the way they are attached to the rachis (figures 19 and 20).

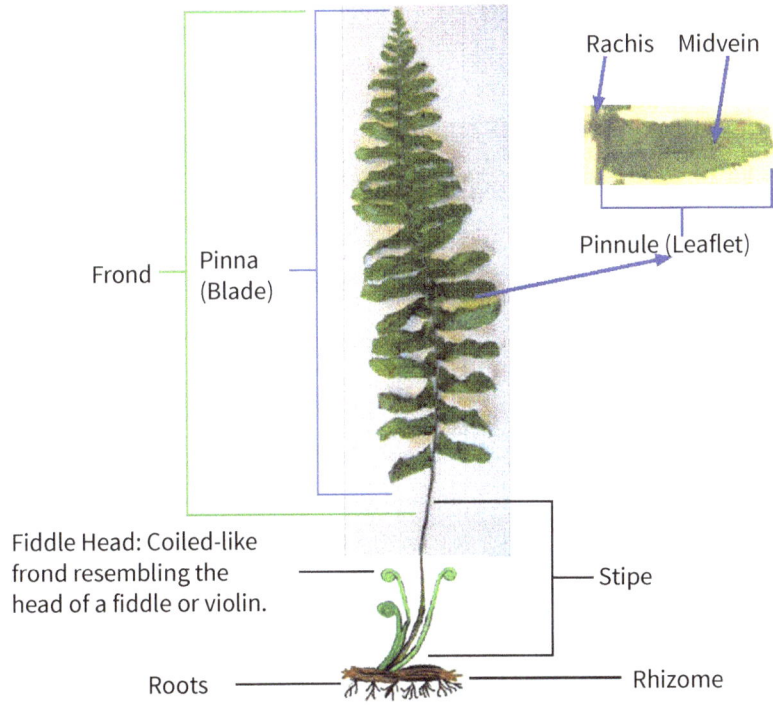

Figure 18. A general illustration of terms used in describing the morphology of a pinnately compound fern frond.

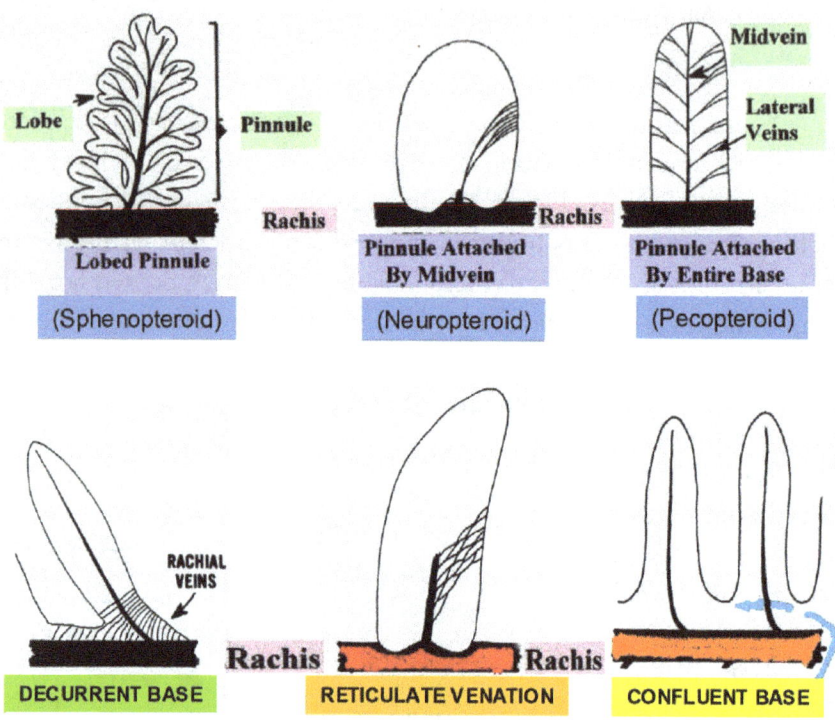

Figure 19. Terms used in describing fern and fern-like foliage. Modified after Gillespie et al., 1978.

Figure 20. A general illustration of terms used in describing the morphology of tripinnate fern fronds from the Phillips coal seam in Wise County, Virginia.

True ferns, or tree ferns, are those that reproduced from spores on the underside of their leaves. Many fragments (or form genera of tree ferns) represent different components of the plant named *Psaronius* (figure 21). The trunk was covered with a wide fibrous root mantle to support the massive 33- to 50-foot-tall plant. Its nearly cylindrical trunk was unbranched, except near the top, where a crown of up to 10-foot-long frond whorls; the fronds were forked 4 to 5 times (compound leaves). *Psaronius* was the dominate plant of the coal swamps (figure

21). The living relatives of the Pennsylvanian age ferns are found in the tropics; these plants are smaller, have shorter stems, and smaller organs than the others. This was found to be true during my fossil collecting expeditions.

The living relatives of the Pennsylvanian age plants are found in the tropics, but they differ from the ancient forms in that they are smaller, have shorter stems, and have smaller organs than their ancient relatives (figures 22 and 23).

Figure 22. A typical tree fern, Cibotium sp., found on the big island of Hawaii. Photo taken by the author in December 2006.

Figure 21. Reconstruction of Psaronius, a tree fern. Modified after Gillespie et al., 1978.

Figure 23. Photograph of a branch of the tree fern Cibotium on the big island of Hawaii.

Plate I. Tree Ferns - *Zygopteridales*. The oldest group of fern fossils. They had their beginning back in the Permian era; see figure 1 for an indication of the time spread between the Pennsylvanian age and the Permian. 1, 2. *Corynepteris angustissima* (Sternberg) preserved in a silty, massive shale. Collected from an outcrop immediately above the Kennedy coal seam, located along U.S. Route 58 Alternate, 1.5 miles east of Coeburn in Wise County, Virginia.

1

2

Herbaceous* ferns were not as common as tree ferns. Perhaps because they were without woody stems, they were not well preserved. Research of the literature did not reveal a reconstruction of the ferns, but they may have appeared somewhat like those shown below in figure 24.

Figure 24. Herbaceous ferns in Hilo on the big island of Hawaii. Phymatosorus grossus (photo on the right).

* Note: Herbaceous refers to a description of a group of plants, including Sphenophyllum, whose stem has little or no woody tissue. They are shrub-like, growing low to the ground and as vines on larger plants.

Plate I. Herbaceous ferns (*Alloiopteris*). 1.Specimen with several blades of Alloiopteris coralloides. 1a. Enlarged view of the lower right corner of *Alloiopteris coralloides*. 2. The morphology of the pinnules. 3. *Alloiopteris coralloides* with multiple fronds attached to the rachis. All specimens are preserved in a sandy shale. Collected immediately above the Kennedy coal seam in an outcrop located along the southbound lane of U.S. Route 58 Alternate, 1.5 miles east of Coeburn in Wise County, Virginia.

4. *Alloiopteris coralloides* preserved in a fine-grained sandstone. Collected from a coal seam rider, 25 feet above the Blair coal seam along I-23 North in Norton, Wise County, Virginia.

Plate II. Herbaceous ferns (*Alloiopteris*). 1. Specimen with several blades of *Alloiopteris coralloides*. 1a. Enlarged view showing the morphology of the pinnules. Specimen preserved in a sandy shale. Collected immediately above the Kennedy coal seam in an outcrop located along the southbound lane of U.S. Route 58 Alternate, 1.5 miles east of Coeburn in Wise County, Virginia.

1

1a

2

3

Plate I. Herbaceous ferns: Alloiopteris.

1

1a

Plate II. Herbaceous ferns: Alloiopteris.

Seed ferns, or pteridosperms, were groups of plants that had large fern-like leaves and reproduced from seeds and pollen-bearing pods (not spores).

Pteridosperms grew as small trees, reaching heights as tall as 10 feet (figure 25), or as shrubs or vines with woody stems. In spite of the shape of the leaves, pteridosperms were more closely related to conifers than to ferns. A modern fern is shown in figure 26.

Pteridosperms were the most common plants of the Pennsylvanian peat marshes. Beginning in the Devonian, the flora thrived into the Mesozoic. Of the hundreds of variations of seed ferns studied worldwide, several of the form genera and species have been found in Virginia. The reproductive organs (seeds and pods), also known as the fructifications, are less common than the foliage of the plants. They are most commonly found separate from the plant stems.

Figure 25. Reconstruction of the seed fern Medullosa. Modified after Gillespie et al., 1978.

Figure 26. The modern "Filmy" fern, Hymenpphyllum, is shown for comparison with Sphenopteris elegans, a stem with multiple fronds preserved in a medium gray shale found in Plate II, Sphenopteris

Plate I. Seed fern - *Neuropteris*. 1. *Neuropteris heterophylla*. 2. *Neuropteris ovata*. Collected from a coal mine in the Parson coal seam along Mud Lick Creek, 2.4 miles northeast of Roda in Wise County, Virginia.

3. *Neuropterocarpus rarinervis* Langford in sandy shale. Collected from an outcrop in the Clintwood coal horizon along I-23 North behind the old Food Lion building of the shopping center located in Wise, Wise County, Virginia.

4. *Neuropteris ovata* in shale. Collected from the roof strata of a mine in the Imboden coal seam on Still House Branch of Roaring Fork, 1.5 miles northeast of Roaring Fork in Wise County, Virginia.

Plate II. Seed fern - *Neuropteris*. 1, 2. *Cyclopteris oblcalaris* preserved in ironstone. Collected from strata immediately above the Blair coal seam in an outcrop along I-23 North in Norton, Wise County, Virginia.

3. A pair of *Neuropteris gigantean* pinnules in shale. Collected from immediately above the Blair coal seam in the Wise Formation along U.S. Route 58 Alternate, approximately 0.5 miles from Appalachia High School in Appalachia, Wise County, Virginia

4. *Cyclopteris* sp. in yellow clay shale. 5, 5a Neuropteris ovata preserved in a sandy carbonaceous shale. Specimens collected from an road cut approximately 2.5 feet below the Lower Split of the Blair coal seam along I-23 North, 1.1 miles from the junction of Route 823 and 5.2 miles south of Pound in Wise County, Virginia.

Plate III. Seed Fern - *Neuropteris*. 1. *Neuropteris oblique* preserved in a very fine-grained carbonaceous sandstone. 2. *Neuropteris oblique* preserved in clay shale. 3, 3a. *Neuropteris oblique* preserved in a medium-grained, reddish-brown ferruginous sandstone. Collected from an outcrop in a road cut along Breaks Park Road, 5.7 miles northeast of Haysi in Buchanan County, Virginia.

Plate IV. Seed fern - *Neuropteris*. 1. *Neuropteris pocahontas* preserved in shale. Collected from strata immediately above the Kennedy coal seam in an outcrop, 0.1 mile east of the junction of U.S. Route 58 Alternate and Boatright Hollow Road in Coeburn, Wise County, Virginia.

2. *Neuropteris* cf. *heterophylla*. 2a. *Neuropteris* cf. *heterophylla* pinnule. Each specimen preserved in a medium-to-dark gray silty shale. Collected immediately above the Kennedy coal seam, 0.1 mile east of the junction of U.S. Route 58 Alternate and Boatright Hollow Road in Coeburn, Wise County, Virginia.

3. *Neuropteris* sp. 4. *Neuropteris* sp. Each specimen preserved in silty shale. Collected immediately above the Phillips coal seam, 4.8 miles northwest of Inman in Wise County, Virginia, along Route 160.

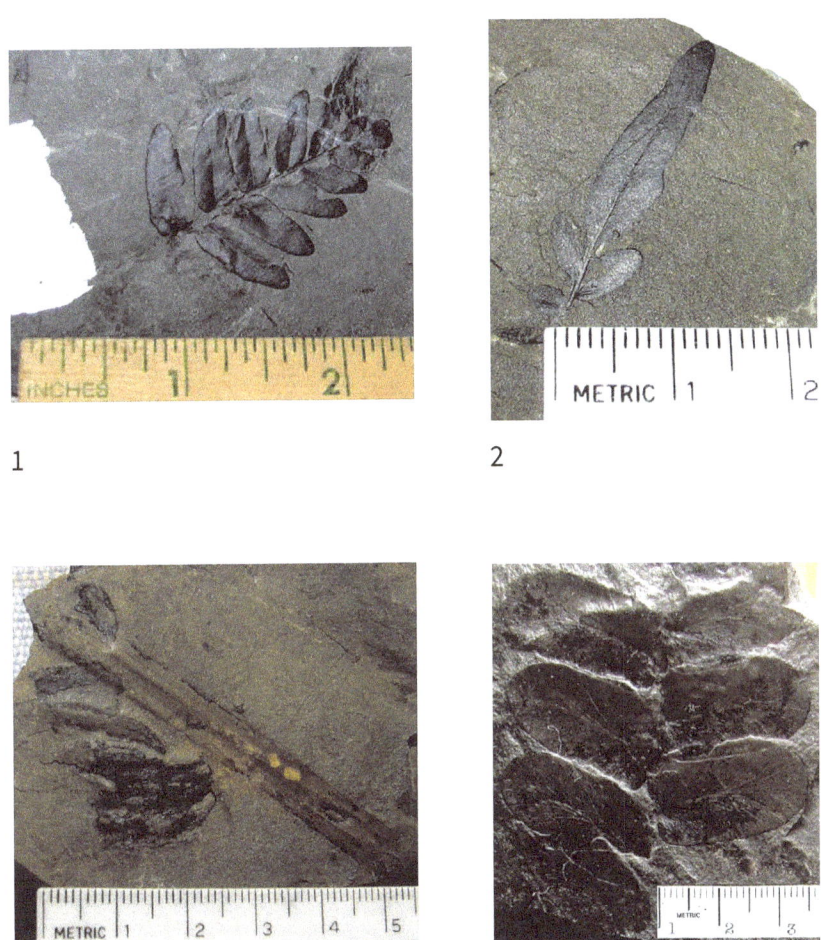

1 2

3 4

Plate I. Seed fern - Neuropteris.

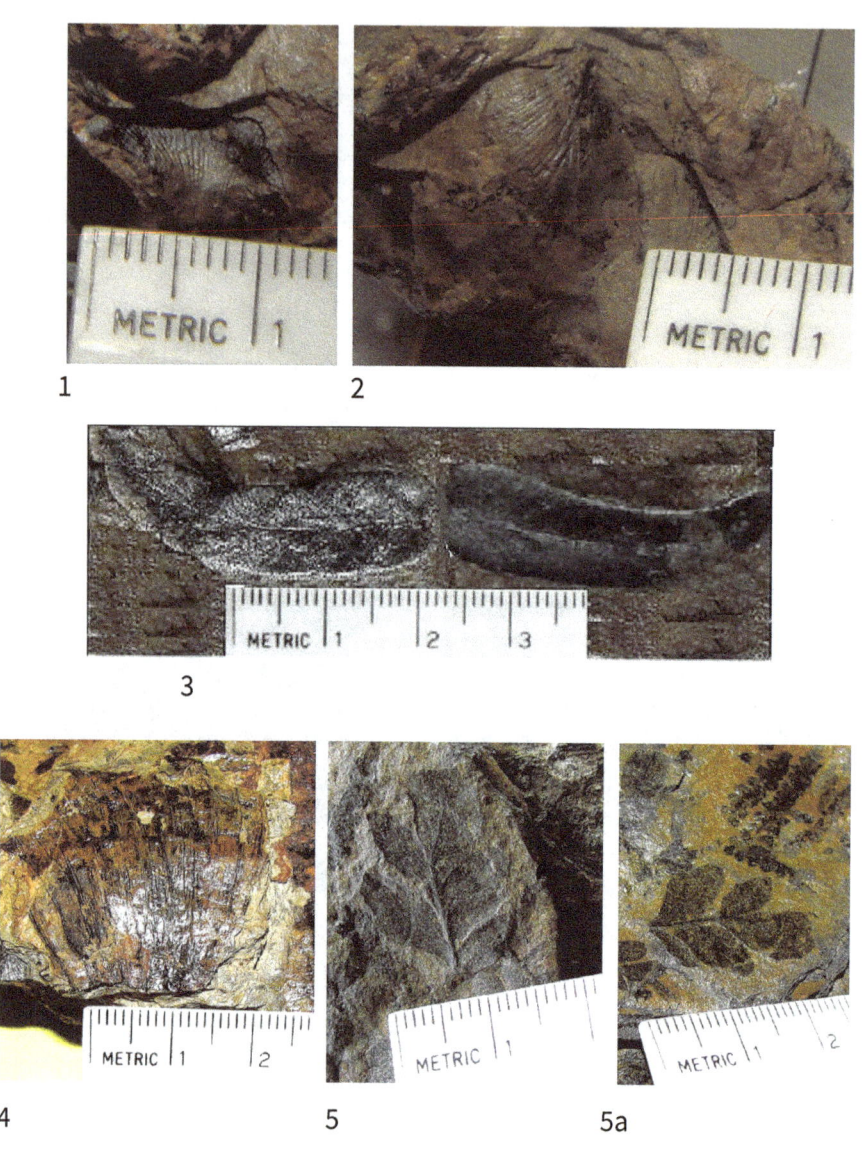

Plate II Seed Fern: Neuropteris

Plate III. Seed fern - Neuropteris.

Plate IV. Seed fern - Neuropteris.

Plate V. Seed fern - *Neuropteris* sp. 1–6. Collected from the Phillips coal seam in an outcrop in a road cut located 4.8 miles northwest of Inman in Wise County, Virginia, on Route 160.

Plate VI. Seed fern - *Neuropteris*. 1 *Neuropteris pocahontas* var. *inaequatis*. Specimen preserved in a light yellow claystone. Specimen preserved in a light yellow claystone. Collected from an outcrop of the Aily coal seam located 1.5 miles west of Coeburn in Wise County, Virginia. The rock primarily exists of the clay mineral montmorillonite, which is derived from volcanic ash.

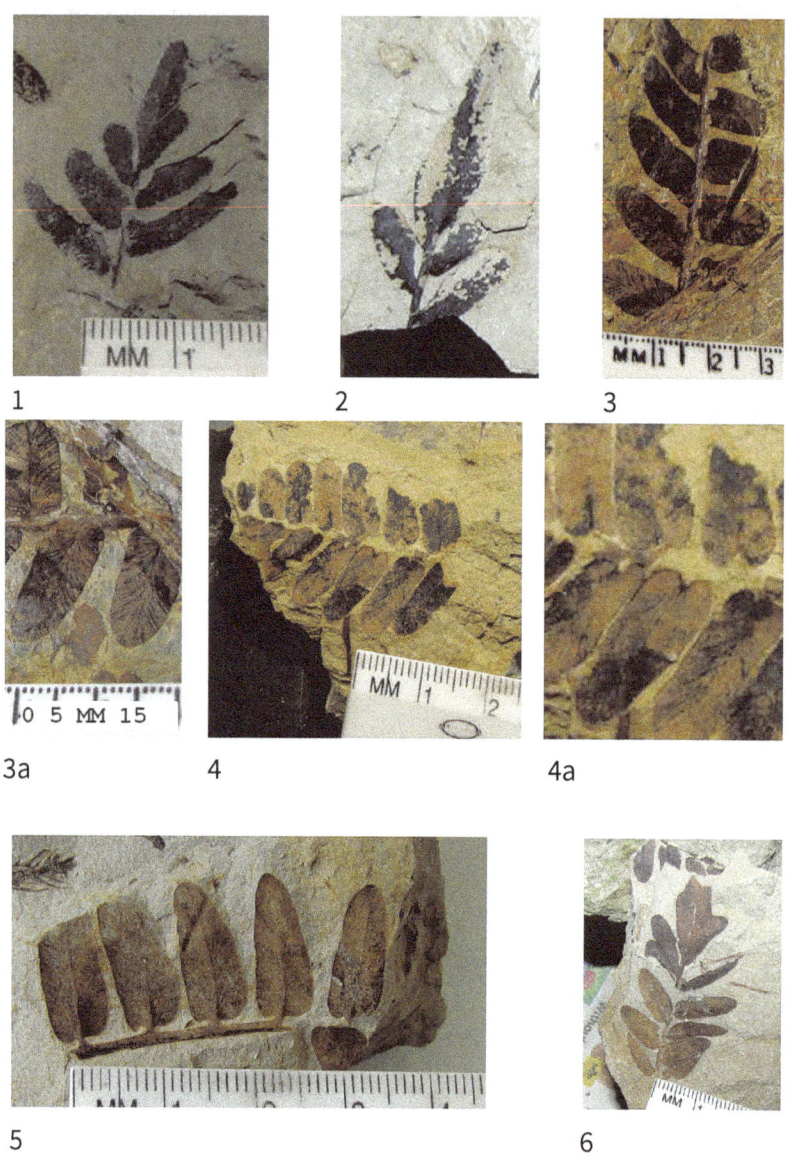

Plate V. Seed fern - Neuropteris.

1

Plate VI. Seed fern - Neuropteris.

Plate I. Seed ferns - *Alethopteris*. 1. *Alethopteris evansi*. Specimen preserved in a light yellow claystone. Collected from an outcrop of the Aily coal seam, 1.5 miles west of Coeburn in Wise County, Virginia. The rock primarily exists of the clay mineral montmorillonite, which is derived from volcanic ash.

2. *Alethopteris serli* in a light gray silty shale. Collected from the Tiller coal seam in Whitewood/Jewell Ridge, Buchanan County, Virginia.

Plate II. Seed ferns - *Alethopteris*. 1, 2. *Alethopteris zeilleri* in two stages of growth. Shown in image 2 is a *calamities* foliage *Lobatannularia* (center). The specimens are preserved in a medium gray silty shale. Collected from the Tiller coal seam in Whitewood/Jewell Ridge, Buchanan County, Virginia.

1

2

Plate I. Seed ferns - Alethopteris.

1

2

Plate II. Seed ferns - Alethopteris.

Plate I. Seed ferns - *Sphenopteris*. 1. *Sphenopteris (Oligocarpia)* sp. in a claystone. Collected from a road cut of the Blair coal seam along I-23 North, 1.1 miles from the junction of Route 823 and 5.2 miles south of Pound in Wise County, Virginia.

2, 2a. *Sphenopteris (Oligocarpia)* sp. in a claystone. 3, 3a. *Sphenopteris pygmaea*. Collected from a road cut of the Glamorgan coal seam horizon along Breaks Park Road, 5.7 miles northeast of Haysi in Buchanan County, Virginia.

Plate II. Seed ferns - *Sphenopteris*. 1. *Sphenopteris elegans* in shale. Collected from the roof strata of a mine in the Jawbone coal seam along Honey Branch Road, just northeast of St. Paul in Wise County, Virginia.

2. *Sphenopteris neuropteroides*. 3. *Sphenopteris pygmaea*. Specimens preserved in gray clay shale. Collected from an Kennedy coal seam road cut near the junction of State Route 158 and U.S. Route 58 Alternate South, about 1.5 miles east of Coeburn in Wise County, Virginia.

4. *Sphenopteris gracilis* in gray shale. Collected from the roof strata of a mine in the Lowsplint coal seam, 2.4 miles north of Stonega on Stonega Road, State Route 78, Wise County, Virginia.

5. *Sphenopteris laurenti* in a gray slickensided shale. Collected from the roof strata of a coal mine in the Upper Banner/Splashdam coal seams along Steels Fork, just northwest of Coeburn in Wise County, Virginia.

Plate III. Seed ferns - *Sphenopteris*. 1. *Sphnopteris striata*. Specimen preserved in a dark gray shale. Collected from the Kennedy coal seam in an outcrop of a road cut located 0.1 mile east of the junction of U.S. Route 58 Alternate and Boatright Hollow Road in Coeburn, Wise County, Virginia.

2, 2a, 2b. *Sphenopteris harveyi* (Leo Lesquereux), 1884, with and without lobes. Specimens preserved in a light gray-to-brownish-white shale. Collected from the Phillips coal seam in an outcrop of a road cut 4.8 miles northwest of Inman in Wise County, Virginia.

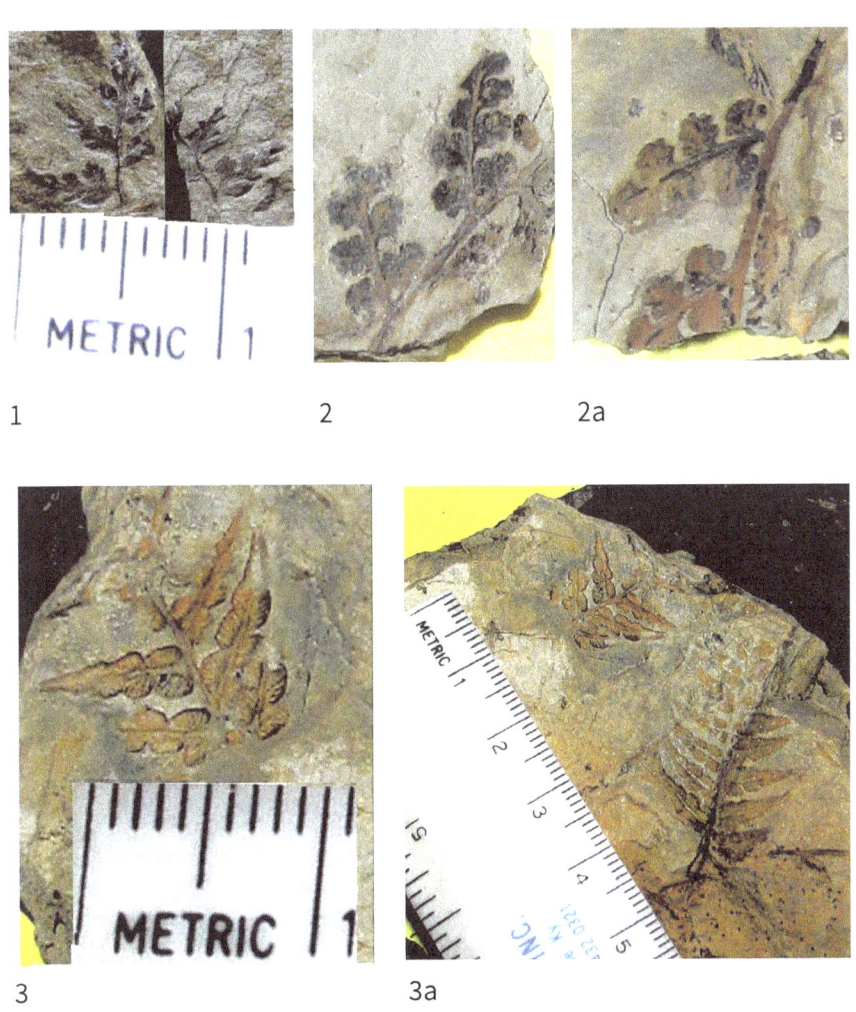

Plate I. Seed ferns - Sphenopteris.

Plate II. Seed ferns - Sphenopteris.

Plate III. Seed ferns - Sphenopteris.

Plate I. Seed ferns - foliage and stem of *Lyginopteris*. 1, 1a. *Sphenopteris spinosa* (Goppert) 1842, complete penultimate pinna. 1b. Enlarged view of a partial ultimate pinna. Collected from the Upper Banner Coal seam in an outcrop of a road cut along the northbound lane of Coeburn Mountain Road in Coeburn, Wise County, Virginia.

2. Stem of *Lyginopteris* in a gray shale. Collected in a mine immediately above the Jawbone coal seam along Mill Branch near Whitewood in Buchanan County, Virginia.

3, 4. Stem of *Lyginopteris* in a gray clay shale. Collected from the shale immediately above the Kennedy coal seam outcrop, 1.5 miles east of Coeburn in Wise County, Virginia, along U.S. Route 58 Alternate.

5. *Diplotheca stellata* is the capsule that surrounded the seed of the *Lyginopteris* plant and is normally found with no seed attached. Specimen preserved in a gray clay shale. Collected from an outcrop of the Phillips coal seam in a road cut 4.8 miles northwest of Inman in Wise County, Virginia, on Route 160 along U.S. Route 58 Alternate.

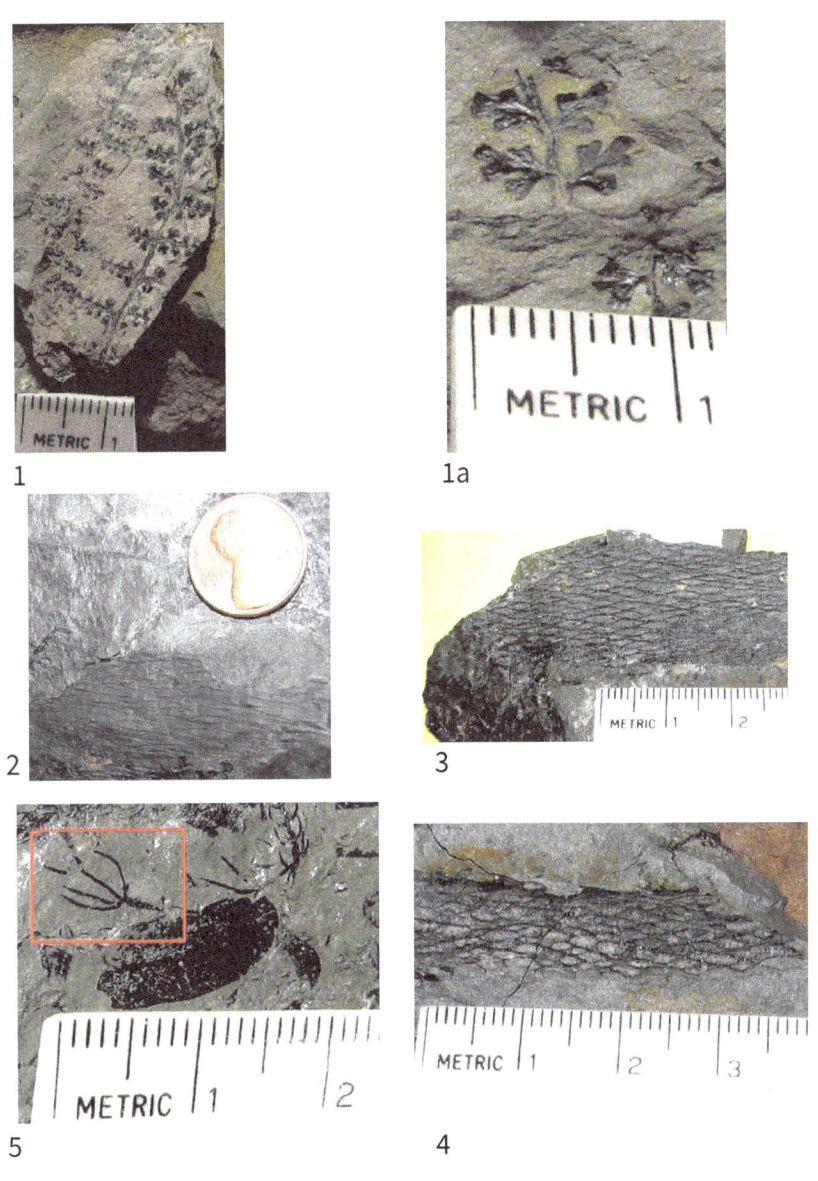

Plate I. Seed ferns - Lyginopteris.

Plate IV. Seed ferns - *Sphenopteris*. 1. *Sphenopteris* sp. 1a. View of specimen in an inverted view, giving the fossil image a 3-D effect. 1b. Greatly enlarged view showing detail. Specimen preserved in a dark gray-to-grayish-black shale. Collected immediately above the Jawbone coal seam in an underground coal mine located on Mill Branch near Whitewood in Buchanan County, Virginia.

Plate I. Seed ferns - *Sphenopteris*.-like. 1, 2. *Zeilleria* sp., "a foliage morphogenus used for forms similar to *Sphenopteris*" (Taylor et al., 2009), in shale. Collected from the Kennedy coal seam in an outcrop of a road cut located 0.1 mile east of the junction of U.S. Route 58 Alternate and Boatright Hollow Road in Coeburn, Wise County, Virginia.

Plate II. Seed ferns - *Sphenopteris*.-like. 1. *Zeilleria* sp. 1a. Enlarged view of one of the blades in figure 1, showing detail of the morphology. Specimen preserved in chlorite shale, which has an unusual silky sheen luster. Collected from the roof strata of a mine in the Splashdam coal seam, located on Deel Fork of Bull Creek at the junction of State Routes 609 and 664, just southwest of Harmon in Buchanan County, Virginia.

Plate I. Seed fern - *Eusphenopteris*. 1, 1a. *Eusphenopteri obtusiloba*. Specimen preserved in a light gray-to-brownish-white shale. Collected from the Phillips coal seam outcrop, 4.8 miles northwest of Inman in Wise County, Virginia, on Route 160.

1 1a

Plate I. Seed ferns - Eusphenopteris.

1a 1b

Plate I. Seed ferns - Sphenopteris.

Plate I. Seed ferns - Sphenopteris-like.

Plate II. Seed ferns - *Sphenopteris*-like.

1

1a

Plate II. Seed ferns - Sphenopteris-like.

Plate I. Seed fern - *Mariopteris*. 1, 1a. 2. Portions of fronds still attached to the rachis. Collected from an outcrop in the Clintwood coal horizon along I-23 North behind the old Food Lion building of the shopping center in Wise, Wise County, Virginia. 3. *Mariopteris anthrapolis* Langford in a sandy shale.

Plate II. Seed fern - *Mariopteris*. 1, 1a, 2. *Mariopteris muricata* preserved in a brownish-yellow silty shale. Collected from the Phillips coal seam in an outcrop in a road cut located 4.8 miles northwest of Inman in Wise County, Virginia, on Route 160.

Plate III. Seed fern - *Mariopteris*. 1. *Mariopteris* sp. preserved in a medium gray silty shale. Collected approximately 25 feet above the Blair coal seam in an outcrop along I-23 North, 0.5 miles from the junction of U.S. Route 58 Alternate in Norton, Wise County, Virginia.

2. *Mariopteris eremopteroides* in a greenish-yellow silty shale. Collected from an outcrop of the Glamorgan coal horizon in a road cut located along Breaks Park Road, 4.5 miles north of the junction of State Routes 76 and 83 near Haysi in Buchanan County, Virginia.

3. *Mariopteris pottsvillea* in a gray shale. Collected from the Phillips coal seam outcrop located 4.8 miles northwest of Inman in Wise County, Virginia, on Route 160.

Plate IV. Seed ferns - *Mariopteris*. 1. *Mariopteris* sp. 1a. View of specimen in an inverted view, giving the fossil image a 3-D effect. 1b Greatly enlarged view showing detail. Specimen preserved in a dark gray-to-grayish-black shale. Collected immediately above the Jawbone coal seam in an underground coal mine located on Mill Branch near Whitewood in Buchanan County, Virginia.

1 1a

2 3

Plate I. Seed fern - Mariopteris.

Plate II. Seed fern - Mariopteris.

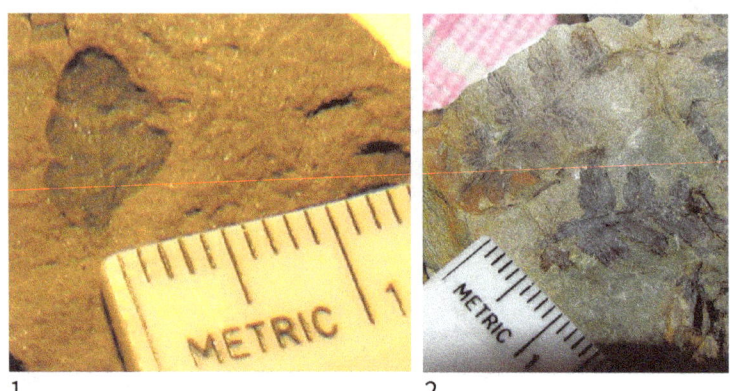

Plate III. Seed fern - Mariopteris.

Plate I. Seed ferns: *Eremopteris*. 1, 1a. *Eremopteris missouriensis* mold and cast. 2, 2a. *Eremopteris crenulata* mold and cast. 3. *Eremopteris* sp. 4. *Eremopteris* sp. (center) with *Sphenopteris harveyi* (Leo Lesquereux), 1884 (right edge of specimen). Specimens preserved in a light brownish-yellow-to-grayish-white silty shale. Collected from just above the Phillips coal seam in a road cut located 4.8 miles west of Inman in Wise County, Virginia, along Route 610.

Plate I. Seed ferns - Eremopteris.

Plate I. Seed fern - *Odonopteris*. 1. *Odonopteris* sp. Specimen preserved in a dark gray shale. Collected from the Kennedy coal seam in an outcrop of a road cut located 0.1 mile east of the junction of U.S. Route 58 Alternate and Boatright Hollow Road in Coeburn, Wise County, Virginia.

2. *Odonopteris aequalis* (Lesquereux), 1866. Mold and cast. 2a, 2b, 2c. Enlargements of leaves for veinalation detail. Specimen preserved in an ironstone concretion. Collected from strata above the Clintwood coal seam horizon in a high-wall at the Wise County Shopping Center along I-23 North, near Wise in Wise County, Virginia.

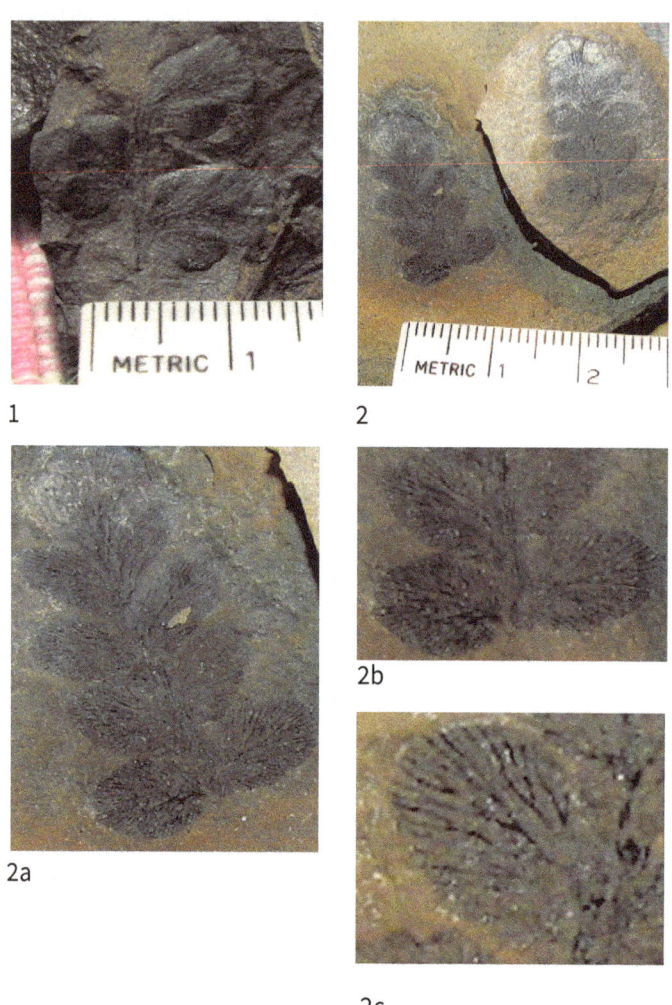

Plate 1. Seed fern - Odonopteris.

Plate I. Seed ferns - *Alethopteris*. 1, 1a. *Alethopteris lonchitica*. 2. *Alethopteris lonchitica* (left) and *Neuropteris* (right). Collected from immediately above the Blair coal seam in the Wise Formation along U.S. Route 58 Alternate, approximately 0.5 miles from Appalachia High School in Appalachia, Wise County, Virginia.

3, 3a. *Alethopteris lonchitica* cast and mold of a nearly complete frond preserved in shale. Collected from the roof strata in a mine located in the Bonnie Blue area, north of St. Charles in Lee County, Virginia.

Plate II. Seed ferns - *Alethopteris*. 1. *Alethopteris decurrens* fronds still attached to rachis in sandy shale. 1a. Enlarged view showing morphology of blades. 2. *Alethopteris serlii* in shale. 3. Several blades of *Alethopteris lonchitica* still attached to the rachis in sandy shale. Specimens collected from the roof strata of a mine in the Parsons coal seam located along Mud Lick Creek, 2.4 miles northeast of Roda in Wise County, Virginia.

Plate I. Seed ferns - Alethopteris.

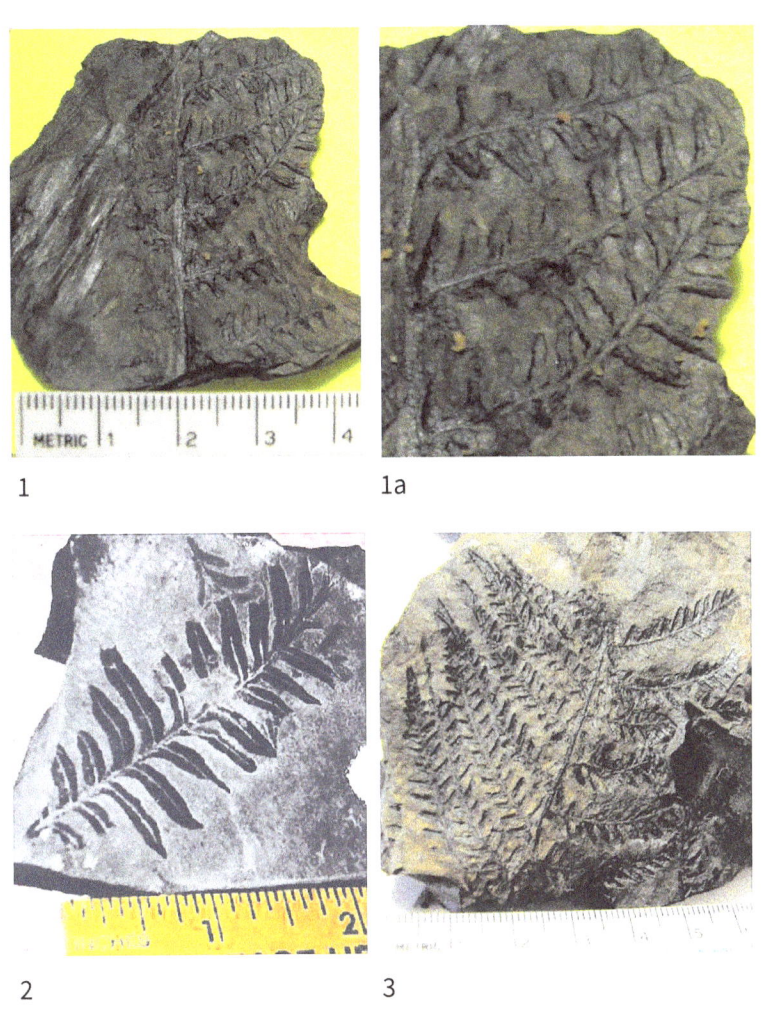

Plate II. Seed ferns - Alethopteris.

Plate III. Seed ferns - *Alethopteris*. 1. *Alethopteris decurrens*. 1a. Enlarged view of a set of pinnules (leaflets) from figure 1. Specimen preserved in a dark gray shale. Collected from the roof strata directly upon the Splashdam coal seam located near Hurley in Buchanan County, Virginia.

1

1a

Plate I. Fern-like stems. 1, 1a, 1b. *Rhodea* sp. preserved in a bluish-gray shale. Collected from the roof strata of a coal mine in the Splashdam coal seam located at the junction of Deel Fork Road and State Route 609 near Harman in Buchanan County, Virginia.

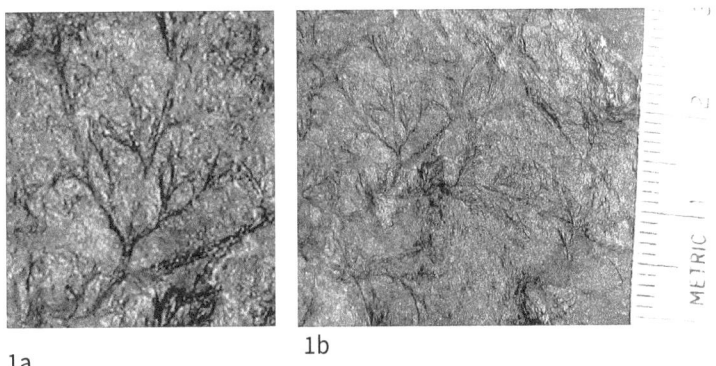

Plate I. Rhodea.

CHAPTER 9
SEEDS

Seeds

Both the pteridosperms and the cordaites produced seeds that could potentially be preserved as fossils. The majority of seeds are generally found separated from the plant that produced them, making it difficult to determine associations, although occasionally seeds are found still attached to the parent plant (see plate II, image 10, and plate I, image 3). Of the seed names (or form groups), *Trigonocarpus* has been associated with the seed fern forms *Alethopteris* and *Neuropteris*. *Neuropterocarpus* is also a seed believed to have been produced by *Neuropteris*. Pollen organs of Medullosa fern (Whittleseya) are represnted in Plate III pollen organs images 1 and 2.

Plate I. Seed pods. 1. *Cardicarpon* sp. or *Crossotheca crepinii*. 2. *Cardicarpon* sp. 3. *Trigonocarpus* sp. nucellus. 4. *Trigonocarpus* sp. *sclerotesta*. 5. *Rhabdocarpus* sp. cast. 5a. *Rhabdocarpus* sp. mold. 6. *Carpolithes*. 7. *Trigonocarpus* sp. 8. *Rhabdocarpus* sp. 9. *Carpolithes*. Specimens 1 and 3–9 collected approximately 2.5 feet below the Blair coal seam along I-23 North, 1.1 miles from the junction of Route 823 and 5.2 miles south of Pound in Wise County, Virginia. Specimen 2 collected from immediately above the Blair coal seam in the Wise Formation along U.S. Route 58 Alternate, approximately 0.5 miles from Appalachia High School in Appalachia, Wise County, Virginia.

10. *Holcospermum* in sandstone. Collected from the Norton Formation immediately above an unnamed coal seam on West U.S. Route 58 Alternate along the railroad tracks in Appalachia, Wise County, Virginia.

11. *Cordaicarpus*. 12. *Trigonocarpus* sp. 13. *Schopfia* sp. Specimens 11–13 collected from the Clintwood coal seam horizon along I-23 North behind the shopping center in Wise, Wise County, Virginia.

Plate II. Seed pods. 1. *Whittleseya* sp. (pollen organ) in gray shale. Collected from the Kennedy coal seam outcrop in a road cut along U.S. Route 58 Alternate East, 0.1 miles east of Boatright Hollow Road near Coeburn in Wise County, Virginia.

2,3. *Cardiocapus bicuspidatus* (Lesquereux), 1884, preserved in a medium gray silty shale. Collected from strata approximately 25 feet above the Clintwood coal seam horizon in a high-wall at the Wise County Shopping Center along I-23 North in Wise, Wise County, Virginia.

4. *Holcospermum* in a light yellow claystone. Collected from the Aily Coal seam outcrop, 200 feet right off U.S. Route 58 Alternate West, 1.5 miles west of Coeburn in Wise County, Virginia.

5. *Carpolithes Butlerianus* (Lesquereux), 1884. 6. *Rhabdocarpus tenax* (Lesquereux), 1884. Specimens preserved in a gray shale. Collected from an outcrop of the Kennedy coal seam in a road cut along 0.1 mile east of the junction of alternate Route 58 and Boaright Hollow Road Coeburn, Wise County, Virginia.

7. *Cardiocarpous*. 8. *Cardiocarpous diplotesta* (Lesquereux), 1884 (Cast and Mold). 9. *Cardiocarpus dilatatus* (Lesquereux), 1884. Specimens preserved in a light gray silty shale to siltstone. Collected from strata approximately 25 feet above the Clintwood coal seam horizon in a highwall at the Wise County Shopping Center along I-23 North, Wise, in Wise County, Virginia.

10. *Neuropterocarpus rarinervis* (Langford) in sandy shale. Specimen Specimens collected from an outcrop in the Clintwood coal horizon along I-23 North behind the old Food Lion building of the shopping center located in Wise, Wise County, Virginia.

Plate I. Pollen organs. 1, 2. *Whittleseya* sp. preserved in a light yellow claystone. Collected from the Aily Coal seam outcrop, 200 feet right off U.S. Route 58 Alternate West, 1.5 miles west of Coeburn in Wise County, Virginia.

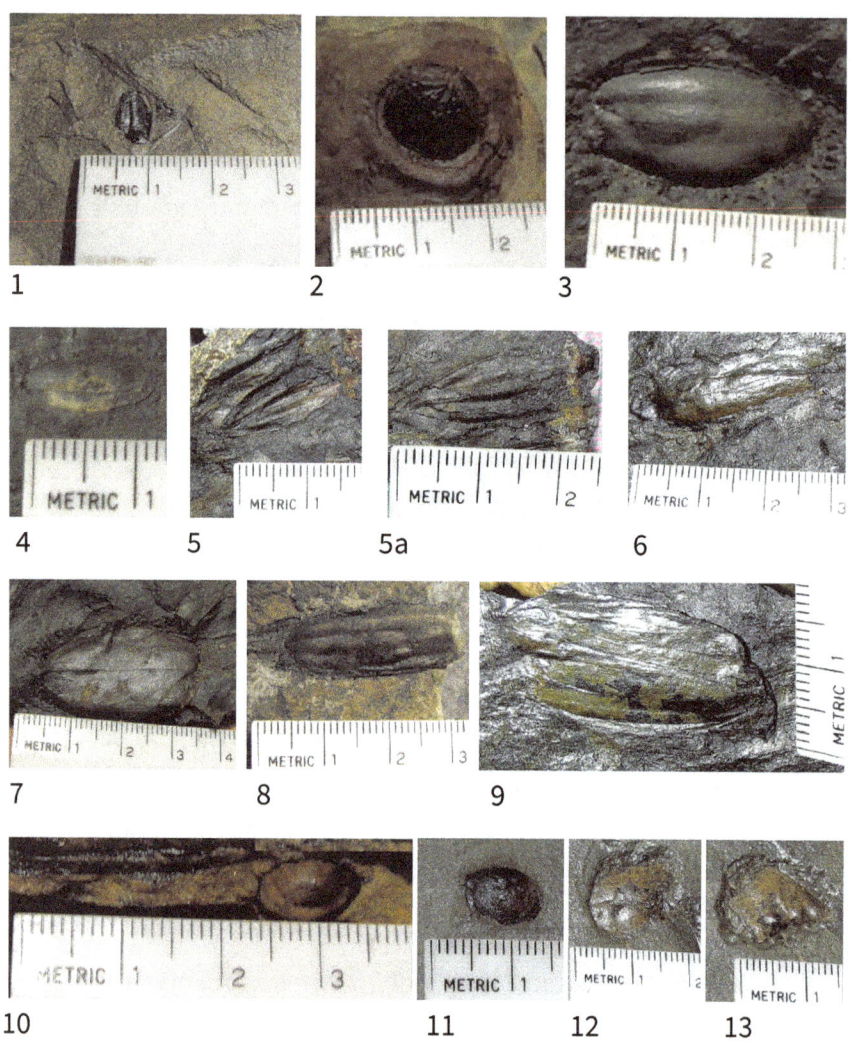

Plate I. Seed pods.

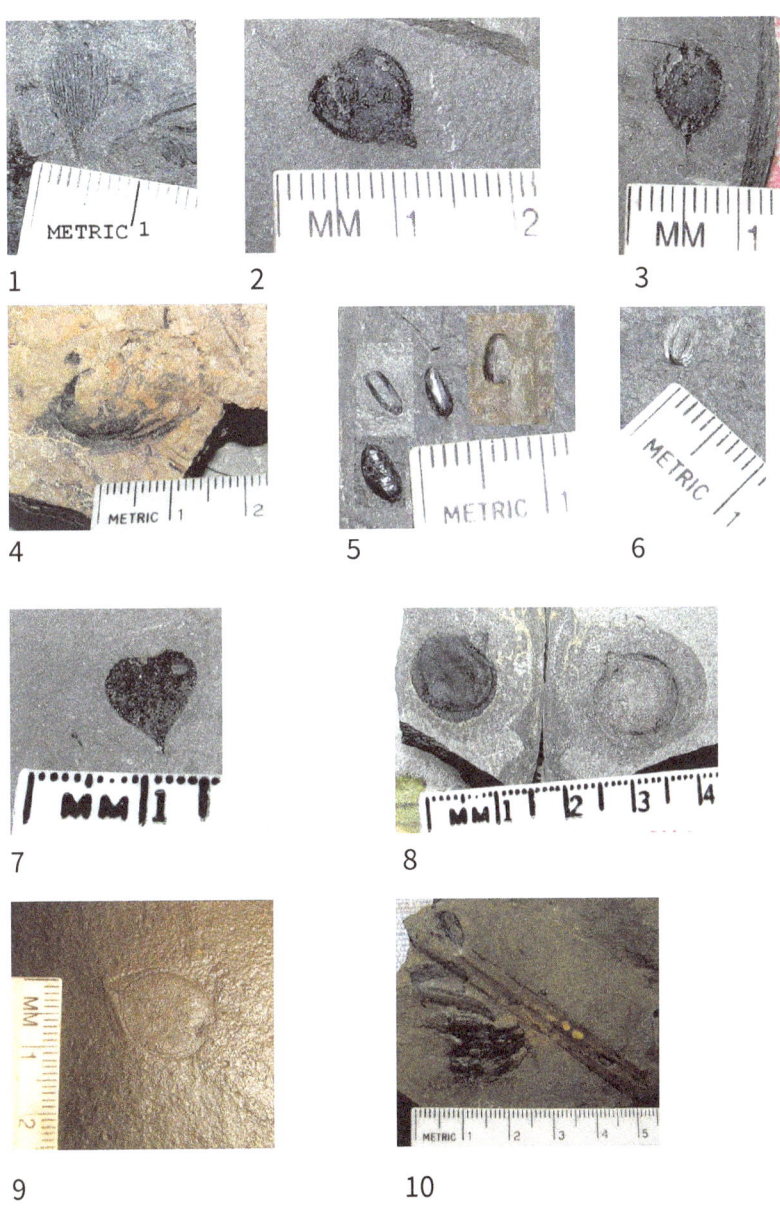

Plate II. Seed pods.

Plate I. Pollen organs

CHAPTER 10

MARINE FOSSIL FAUNA

FOUND WITH PLANT FOSSIL FLORA

This chapter presents several types of ocean and near-shore dwelling life forms (fauna) that lived adjacent to or migrated into marsh areas after the vegetation (flora) were drowned and buried under sediments. These sediments became the roof rocks in coal mines. The fossil fauna shown in this chapter were collected just above or within a few feet of a few of the coal seams listed in figure 1.

The groups of fauna that possessed an exterior shell-like skeleton consisting of a bottom shell and a top shell (shown in plates I and II of this chapter) are known as "bivalve." These are the brachiopods. They come both with and without hinges (joints of articulation). The majority of these creatures are extinct except for one, the *lingula*. Note that the name *lingula* comes from Latin and means "little tongue." These bivalved creatures are not to be confused with the Pelecypods or "clams" represented here as the *Pecinacea, pecten* (see plate I, image 1). These life forms are still living today and are called scallops.

Another type of fauna with an external skeleton or shell that is "flat-coiled" is one of the cephalopods, including the nautiloids (see plate IV, image 2). The last type of fauna that was found is a creature that has an internal, cone-shaped shell (see plate IV, images 1, 1a, and 3). These are squid-like marine organisms.

Geologists use certain species of the faunal groups described above to determine the environment(s) in which they lived. Also, certain members of each of the groups of organisms are certain specific forms that are used as a guide for dating the age of the associated rocks, coals, and fossilized plants. These are the ones which existed only during short periods of geologic time and are referred to as "guide fossils" (Moore, 1952, pp. 197 and 335). It has been found by geologists that the *lingula, pecten* and *nautilus* are found both in ancient rocks as fossils and in direct descendants living in the modern seas (called "living fossils").

Plate I. Marine fossils - pelecypods. Dysodont types of *Pectinacea*. Of the Carboniferous pelecypods, these forms were dominant. 1. *Pecten* sp. 2. *Aviculopecten* cast. 3. *Aviculopecten* mold. 4, 5. *Fasciculiconcha* mold. 6. *Cordaicarpus* (Geinitz) Stewart, 1917 (seed pod), found with two incomplete pectens. The shells are preserved in a light gray siltstone. Collected from an outcrop in the Clintwood coal horizon along I-23 North behind the shopping center in Wise, Wise County, Virginia.

Plate II. Marine fossils - pelecypods. 1. *Chaenomya*. 2. *Mytilarca*. Collected from the Norton coal horizon, 0.6 miles west of the junction of State Routes 817 and 637 on Bold Camp Mountain, 2 miles south of Pound in Wise County, Virginia.

Plate III. Brackish water fossils - brachapods. 1. *Lingula*. 1a. Modern living *Lingula*. 2. *Lingulacea* sp. (Note that the preservation of the original shell material is mother-of-pearl.) Collected from the Norton coal seam horizon along Thackers Branch Road in the Dorchester community of Norton in Wise County, Virginia.

Plate IV. Marine fossils. 1, 1a. Mold and cast of a nautiloid, possibly *Bactrites or Psudeoorthoceras*. 2. Nautiloid, possibly *Parametacoceras* sp. 2a. Modern living *Nautilus*. The shells are preserved in a light gray siltstone. Collected from an outcrop in the Clintwood coal horizon along I-23 North behind the shopping center in Wise, Wise County, Virginia.

Plate I. Marine fossils (Peleceypods).

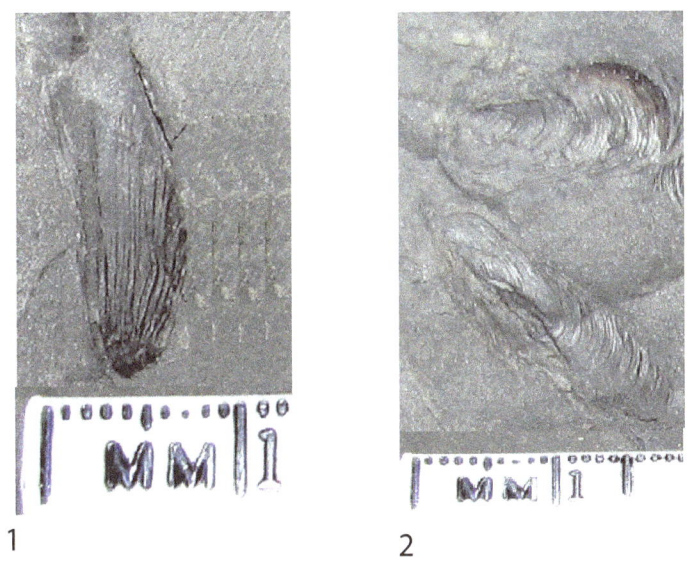

Plate II. Marine fauna (Pelecypods).

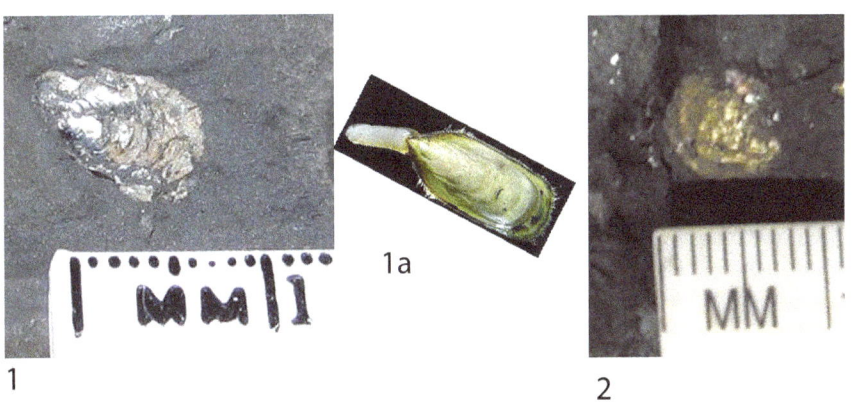

Plate III. Brackish water fauna (Brachiopods)

Plate IV. Marine fauna (Nautiloid).

REFERENCES:

Case, Gerald R. 1982. *A Pictorial Guide to Fossils*. New York: Van Nostrand Reinhold Company.

Chase, Frank E. and Gary P. Sames. 1983. "Kettlebottoms: Their Relation to Mine Roof and Support." U.S.Bureau of Mines RI 8785:12.

Dilcher, D. L. and T. A. Tott. 2005. "Atlas of Union Chapel Mine Fossil Plants." In *Pennsylvanian Footprints in the Black Warrior Basin of Alabama*," edited by R. J. Buta, A. K. Rindsberg, and D. C. Kopaska-Merkel. Alabama Paleontological Society Monograph, no. 1: 339–365.

Fortey, Richard. 1982. *Fossils: The Key to the Past*. New York: Van Nostrand Reinhold Company.

Francis, W. 1961. *Coal: Its Formation and Composition*. 2nd edition. London: Metcalfe & Cooper Ltd.

Gillespie, William H., John A. Clendening, and Herman W. Pfefferkorn. 1978. "Plant Fossils of West Virginia." *West Virginia Geological and Economic Survey*, Education Series ED-3X: 172.

Janssen, Raymond, E.1939. "Leaves and stems from fossil forests: A handbook of the paleobotanical collections of the Illinois State Museum." *Popular Science*, Series Vol 1: 190.

Kukuk, Paul. 1938. *Geologie des iederrheinisch-Westfalischen*.

Lesquereux, Leo. "1879–1884: Description of the coal flora of the Carboniferous formation in Pennsylvania and throughout the United States." *2nd Pennsylvania Geological Survey Publication*, 3 volumes.

Matthews III, William H. 1962. *Fossils: An Introduction to Prehistoric Life*. New York: Barnes & Noble.

Moore, Raymond, C. et al. 1952. *Invertebrate Fossils*. New York: McGraw-Hill.

Ohio Department of Natural Resources, Division of Geological Survey.1996. *Fossils of Ohio*, Bulletin 70: 577.

Read, Charles B. and Serguis H. Mamay. 1964. "Upper Paleozoic Floral Zones and Floral Provinces of the United States." *United States Geological Survey*. Professional Paper 454-K: 79.

Brousmiche, C. 1983. "Les Fougéres sphénoptéridiennes du Bassin Houiller Sarro-Lorrain." Société Géologique du Nord: 10.

Seward, A. C. 1898. *Fossil Plants: A text-book for students of Botany and Geology*. Volumes I, II, and III. London: Cambridge University Press.

Steinkohlengebietes. New York: Springer-Verlag.

Taylor, Thomas, N., Edith L. Taylor, and Michael Krings. 2009. *Paleobotany: The Biology and Evolution of Fossil Plants*. 2nd edition.

Thomas, B. A. 1968. "The carboniferous fossil lycopod *Ulodendron landsburgii* (Kidston) comb. nov." *Journal of Natural History*: 425–428.

Tidwell, William D. 1998. *Common Fossil Plants of Western North America*. 2nd edition. Washington DC: Smithsonian Institution Press.

APPENDIX A

Locations of Fossil Collection Sites in Virginia

7.5 Minute Topographic Quadrangle Coal Map	Seam	Longitude*	Latitude *	Town/City	Geologic Formation +
Hurley	Splashdam1**	82-04-30	37-25-30	Hurley	Norton
Grundy	Splashdam2	82-30-00	37-20-00	Hurley	Norton
Harman	Splashdam3	82-13-00	37-17-00	Harman	Norton
Harman	Splashdam4	82-14-00	37-19-00	Conaway	Norton
St. Paul	Jawbone1	82-19-00	36-57-00	St. Paul	Norton
Carbo	Jawbone2	82-10-30	36-59-30	South Clinchfield	Norton
Bradshaw	Jawbone3	81-50-00	37-15-00	Whitewood	Norton
Appalachia	Parsons (Pardee)	82-49-17	36-58-13	Roda	Wise
Coeburn	Kennedy1	82-24-00	36-56-15	Coeburn	Norton
Coeburn	Kennedy2	82-24-59	36-56-38	Coeburn	Norton
Nora	Upper Banner1	82-15-15	37-05-30	Counts	Norton
Appalachia	Taggart	82-47-30	37-00-00	Bluff Spur	Wise
Appalachia	Blair1	82-46-06	36-54-45	Appalachia	Norton
Pound	Blair2	82-35-15	37-01-15	Pound	Wise
Wise	Blair3	82-36-45	36-56-15	Norton	Wise
Wise	Blair4	82-35-15	36-57-40	Wise	Wise
Appalachia	Unnamed	82-47-30	36-53-15	Appalachia	Norton
Honaker	Tiller	81-52-35	37-07-30	Red Ash	Norton
Appalachia	Phillips	82-50-41	36-55-23	Inman	Wise
Wise	Aily	82-30-50	36-56-11	Tacoma	Norton
Nora/Duty	Upper Banner2	82-15-15	37-03-00	Bucu	Norton
Coeburn	Upper Banner/ Splashdam	82-29-55	36-57-56	Coeburn	Norton

APPENDIX A

(continued)
Locations of Fossil Collection Sites in Virginia

7.5 Minute Topographic Quadrangle Coal Map	Seam	Longitude*	Latitude *	Town/City	Geologic Formation +
Pennington Gap	Harlan	83-02-00	36-51-00	St. Charles	Wise
Norton	Blair5	82-36-46	36-56-30	Norton	Wise
Appalachia	Lowsplint	82-47-22	36-57-40	Stonega	Wise
Coeburn	Upper Banner3	82-29-25	36-56-53	Coeburn	Norton
Carbo	Lower Banner	82-11-52	36-59-21	South Clinchfield	Norton
Vansant	Hagy	82-08-15	37-11-23	Vansant	Norton
Elkhorn City	Glamorgan	82-16-47	37-15-52	Haysi	Wise
Pound	Norton	82-36-13	37-07-27	Pound	Norton
Norton	Clintwood	82-35-50	36-58-18	Wise	Wise
Flat Gap	Low Splint	82-40-10	37-03-45	Norton	Norton
St. Paul	Upper Banner4	82-15-59.95	37-00-01	Dante	Norton
Flat Gap	Low Splint	82-40-10	37-03-45	Norton	Middle Wise
St. Paul	Upper Banner4**	82-15-59.95	37-00-01	Dante	Norton

Footnotes: * Coordinates are listed in degrees-minutes-seconds.
* Number designates a different location but same coal seam horizon.
+ Geologic age is Upper Lower to Upper Pennsylvanian.

CPSIA information can be obtained
at www.ICGtesting.com
Printed in the USA
LVHW06s1621240918
591194LV00016B/669/P